SELECTED READINGS IN PHYSICS
General Editor: D. TER HAAR

THE SOLID STATE MASER

THE SOLID STATE MASER

BY

J. W. ORTON

D. H. PAXMAN

J. C. WALLING

Solid State Physics Division
Mullard Research Laboratories

PERGAMON PRESS

Oxford · New York · Toronto
Sydney · Braunschweig

PERGAMON PRESS LTD.,
Headington Hill Hall, Oxford

PERGAMON PRESS INC.,
Maxwell House, Fairview Park, Elmsford, New York 10523

PERGAMON OF CANADA LTD.,
207 Queen's Quay West, Toronto 1

PERGAMON PRESS (AUST.) PTY. LTD.,
19a Boundary Street, Rushcutters Bay, N.S.W. 2011, Australia

VIEWEG & SOHN GMBH,
Burgplatz 1, Braunschweig

First edition 1970

Library of Congress Catalog Card No. 74-101374

Printed in Great Britain by Bell and Bain Ltd., Glasgow.

08 006818 9 (flexicover)
08 006819 7 (hard cover

Contents

Preface

IN THIS volume we have attempted to tell the story of the solid state maser amplifier from the first tentative theoretical proposals which appeared in the early 1950s to the successful realisation of practical devices and their application to satellite communications and radio astronomy almost exactly ten years later. In concentrating on this particular aspect of maser development, we have done far less than justice to gaseous masers, and, the ammonia maser in particular, which not only preceded the solid state maser in time, but has a history of its own (with considerable potential as a frequency standard). We offer our apologies to the reader (and to those concerned in research into gas masers!) for the distortion inevitably introduced by our giving "only one side of the story", but the space available to us and our own interests (and qualifications!) persuaded us this was the only course possible.

In Chapter 1 we present an historical account of the early developments (including that of the ammonia maser) then digress in Chapter 2 to sketch in a certain amount of necessary background on the properties of paramagnetic ions in crystals before describing the development of practical low noise amplifiers in Chapter 3. Chapter 4 describes some of the characteristics of maser devices designed for communications use. Part II contains reprints of several important papers which illustrate various aspects of the subject. The reader who has followed the introduction in Part I should not experience too much difficulty in understanding them with, perhaps, the exception of those using the spin Hamiltonian method for describing paramagnetic energy levels. We have therefore included in Chapter 2 a short summary of the appropriate theoretical background which should provide the necessary clarification.

PART I

CHAPTER 1

An Historical Account of Early Developments

1.1. Introduction

For many years physicists have been interested in the interaction of radiation and matter and some understanding of this subject is necessary to appreciate the line of development which led to the production of the microwave maser amplifier. Many of the basic ideas required were developed in the study of atomic spectroscopy where the interaction of light waves and atomic systems was first investigated (see, for example, Hindmarsh, 1967). Such studies led to the realisation that these systems exhibit a series of energy levels, between which radiation may induce transitions when the frequency of the radiation satisfies the condition:

$$h\nu_{12} = E_1 - E_2 \tag{1.1}$$

h is Planck's constant and E_1 and E_2 are the energies of the levels involved. For the visible region of the spectrum the frequency ν_{12} is of the order of 10^{15} Hz. As the study of these spectra developed and the resolution of the apparatus increased attention focused on the detailed structure of many of the spectral lines caused by fine splittings of the energy levels involved. The smallest splittings observable by optical techniques correspond to frequencies of between 10^{10} and 10^{12} Hz.

The technical advances in radio and radar during World War II provided physicists with a whole range of sources of radiation in the frequency region 10^6 to 10^{10} Hz. These sources were powerful oscillators providing a few milliwatts of coherent monochromatic radiation which could often be tuned over a small range. From

1945 onwards there was intense activity in the utilisation of these sources to investigate the fine splitting of molecular and atomic energy levels directly. In particular, advances were being made simultaneously in three fields of interest in this discussion, namely gaseous spectroscopy, nuclear magnetic resonance and electron paramagnetic resonance. In the first case the energy levels of interest were the fine splittings of the vibrational–rotational levels produced by certain types of symmetry within the molecule; such splittings correspond to frequencies of the order of 3×10^{10} Hz or microwaves of 1 cm wavelength. In the second case the splittings arise from the interaction of the magnetic moment of an atomic nucleus with a magnetic field. These separations are of the order of $\mu_N H$ where μ_N is the nuclear magnetic moment and H is the applied magnetic field. For a field of 1 kilogauss the level separations produced are of the order of 3×10^6 Hz, a typical radio frequency. In the third case the splittings studied were the well-known Zeeman splittings produced by the action of an applied field on the unpaired electron spins of paramagnetic ions. Here the frequencies involved are approximately 10^3 times as great as in the nuclear case due to the larger magnetic moment of the electron, so they correspond to frequencies of the order of 3×10^9 Hz, a typical microwave frequency.

The experiments performed in these three branches of spectroscopy were very similar. A resonant circuit was often used (a coil or microwave cavity) to concentrate the radiation field around the specimen which was placed inside. The frequency of the oscillator was varied or the magnetic field adjusted to "tune" the levels until the condition given in equation (1.1) was satisfied for a pair of levels within the ion or molecule, when a resonant absorption could occur. In view of the similarity of the experiments, many of the techniques and theories developed in one of these areas were readily adapted for use in the other two.

As early as 1917 Einstein (1917) had considered such interactions and had discussed the way in which radiation could cause transitions between the energy levels of atomic systems, in particular drawing attention to the concepts of stimulated absorption and emission which underlie the whole process of maser action. A treatment using a quantum mechanical approach by Dirac (1927) led to similar results. The following simple discussion illustrates the main points. (For further discussion see, for example, ter Haar (1967).)

Consider the system of two energy levels of energy E_1 and E_2 ($E_2 > E_1$) in equilibrium with the radiation inside a black-body enclosure at temperature T. The radiant energy density at the frequency

$$\nu_{12} = \frac{(E_2 - E_1)}{h}$$

is given by the Planck radiation law:

$$\rho_0(\nu_{12}) = \frac{8\pi h \nu_{12}^3}{c^3}\left\{\frac{1}{\exp(h\nu_{12}/kT) - 1}\right\} \tag{1.2}$$

Let B_{12} be the probability per unit time for unit radiation density that an atom in level 1 will make a transition to the upper level 2 (absorbing energy from the field) and let B_{21} be the corresponding downward probability. It is also necessary to consider the possibility that atoms in the upper level may undergo downward transitions even in the absence of the field. This "spontaneous emission" probability is denoted by A_{21}.

Let n_1 and n_2 be the populations of the levels which at thermal equilibrium are related by the well-known Boltzmann law:

$$\frac{n_1}{n_2} = \exp\left(\frac{h\nu_{12}}{kT}\right) \tag{1.3}$$

Also, at equilibrium the upward and downward transition rates must be equal so we have:

$$n_1 B_{12} \rho_0(\nu_{12}) = n_2 B_{21} \rho_0(\nu_{12}) + n_2 A_{21} \tag{1.4}$$

In the limit of infinitely high temperature equation (1.3) shows that $n_1 = n_2$ and equation (1.2) shows $\rho_0(\nu_{12})$ to become very large. Putting these two conditions in equation (1.4) then yields the relation:

$$B_{12} = B_{21} \tag{1.5}$$

Combining (1.3) and (1.4), using (1.5) and solving for $\rho_0(\nu_{12})$, gives

$$\rho_0(\nu_{12}) = \frac{A_{21}}{B_{21}}\left\{\frac{1}{\exp(h\nu_{12}/kT)-1}\right\} \tag{1.6}$$

which, when compared with the Planck radiation law (1.2), gives the relationship between A_{21} and B_{12}:

$$A_{21} = \frac{8\pi h\nu_{12}^3}{c^3} B_{12} \tag{1.7}$$

The process by which the atomic system absorbs energy from the radiation field is known as stimulated absorption and its reverse as stimulated emission, as in both cases the transition probability is proportional to the radiation intensity. If we now suppose the system subjected to an applied radiation field at ν_{12}, stimulated emission will exceed spontaneous emission when $B_{12}\rho(\nu_{12}) > A_{21}$, i.e.

$$\rho(\nu_{12}) > \frac{8\pi h\nu_{12}^3}{c^3} \tag{1.8}$$

For microwaves of frequency 3×10^{10} Hz (1 cm wavelength) condition (1.8) gives:

$$\rho(\nu_{12}) > 8\pi\times6{\cdot}6\times10^{-34}\ \text{joules/cm}^3$$

corresponding to a power level of approximately 10^{-22} watts in a waveguide of 1 cm^2 cross-section. Thus with the powerful microwave sources available, spontaneous emission can be neglected and the interaction can be characterised by the stimulated transition probability $B_{12}\,\rho(\nu_{12})$ for upward and downward transitions. It is of interest to note that, excluding the use of lasers, condition (1.8) cannot be satisfied at optical frequencies so there the effects of the spontaneous emission are important.

Considering experiments performed at microwave or radio frequencies radiation will stimulate transitions between the two levels E_1 and E_2 with a *net* upward rate of $B_{12}\,\rho(\nu_{12})\,(n_1-n_2)$, where n_1 and n_2 are the numbers of ions in the two levels.

At thermal equilibrium $n_1 > n_2$ as given by the Boltzmann law (equation (1.3)) and, as each upward transition involves the absorption of a quantum of energy $h\nu_{12}$ from the field, there will be a net power absorption:

$$P_{\text{abs}} = W_{12}(n_1 - n_2)h\nu_{12} \tag{1.9}$$

where $W_{12} = B_{12}\,\rho(\nu_{12})$ is the transition probability per ion per unit time for the transition $1 \rightarrow 2$. This absorption process tends to reduce the population difference $(n_1 - n_2)$, but is opposed by thermalising processes which try to maintain thermal equilibrium. These thermalising or relaxation processes may occur through interatomic collisions in the case of gases or interaction with lattice vibrations in solids.

If the net rate of upward transitions induced can be made very much faster than the rate at which the thermalising processes restore thermal equilibrium, it is possible to change the populations n_1 and n_2 and in some cases a condition of saturation where $n_1 = n_2$ can be achieved.

Early in the 1950s physicists began to realise that not only was it possible to change the populations in the energy levels, but it might be feasible to make $n_2 > n_1$. In that case equation (1.9) indicates that instead of absorbing power the system would be capable of emitting radiation to a stimulating field at a rate proportional to the intensity of the radiation field. It can be seen from the above remarks that such an "inverted population" would be maintained only for a short time depending on the rate of the thermalising processes unless some means of maintaining the inversion could be devised.

In the case of nuclear magnetic resonance these relaxation processes often have a time constant as long as a few minutes so such a reversed population could be readily detected. First Bloch (1946) and later Purcell and Pound (1951) obtained a reversal of a spin population which lasted for a short period. The latter workers used a crystal of LiF in which the Li^7 nuclei have a magnetic moment. Their measurements of relaxation times showed that at high fields (6376 gauss) the relaxation time was 300 sec, falling to 15 sec at zero field. In the inversion experiment the LiF was placed in the high field and allowed to assume equilibrium. It was then quickly removed to a second small coil in which the field (100 gauss) was rapidly reversed after which the specimen was returned to the main coil. They found that the Li resonance signal was now inverted but it decayed back to the normal equilibrium condition with a 300-sec time constant. The reversed magnetisation lasted for about 1·75 min. Purcell and Pound used the term "negative temperature" to describe the inverted population, this term arising from the use of the Boltzmann factor with T now negative to indicate that $n_2 > n_1$.

In 1953 Weber (1953) suggested that such inverted populations could be used in an amplifying device, pointing out the possibility of pulse amplification from systems with inverted magnetic ion populations and he also considered the use of a gas to obtain continuous operation; in this case the molecules were to be passed in a stream through a region of reversed electric field. It was this latter suggestion which bore fruit for in 1954 and 1955 Gordon, Zeiger and Townes (1954, 1955) published papers describing the continuous operation of an oscillating device working with an inverted population of two of the levels of the ammonia molecule. Because the ammonia molecule can exist in two forms with the nitrogen atom on either side of the plane of the three hydrogen atoms, all the rotational–vibrational levels of the molecule are split into two, the separation between the lowest levels corresponding to a frequency of 24×10^9 Hz. As the two states of the molecule have slightly different behaviour in electrostatic fields it is possible to effect a separation of the two types of molecule by passing a stream of ammonia gas through a suitable electrostatic lens system (Stern and Gerlach (1922) had performed a very similar magnetic state separation in 1922). In the device described by Gordon, Zeiger and Townes the separation system was designed to focus molecules in the upper energy state into a microwave cavity tuned to the transition frequency between the two states. The molecules in the cavity reverted to the lower energy state due to the stimulation effects of radiation present as noise in the cavity; these molecules emitted radiation which in turn caused more stimulated transitions, and if the molecular beam was sufficiently intense the cavity losses could be overcome and the device would oscillate continuously. The frequency of the oscillation was fixed by the properties of the ammonia molecule at 23,870 MHz. As the gas pressure had to be kept low in order to minimise collision processes by which the equilibrium population could be restored, the power output was only 10^{-9} watts, and the device found most use as a microwave frequency standard. The term Maser (**M**icrowave **A**mplification by **S**timulated **E**mission of **R**adiation) was first used in connection with this device.

The use of a paramagnetic material in place of a gas in such a system offered the possibility of tuning the frequency with an external magnetic field and of increased power output if the concentration of the active ions could be increased. An attempt was made by Combrisson, Honig and Townes (1956) to obtain oscillations at 9000 MHz using phosphorous donors in a silicon host lattice. Inversion of the two levels was obtained by the adiabatic fast-passage technique, where the applied magnetic field is swept rapidly through the resonant condition (Pake, 1962). Although the relaxation time of the phosphorous donors was between 5 and 30 sec at 2°K, the concentration of atoms used was insufficient to produce oscillations within the cavity. They did point out in the conclusion of their paper that due to the low operating temperature, 2°K, the noise temperature of such paramagnetic devices should be low. This appears to be the first reference to this particular property of the solid state maser which makes it outstanding as an amplifying element.

Basov and Prokhorov (1955) made a further step forward by suggesting a different mechanism by which the necessary population inversion could be maintained in a system of gas molecules. As mentioned above, if a sufficiently intense radiation field is applied to a pair of energy levels in an atomic system, it is possible to cause transitions at such a rate that the two populations become equal. The power necessary to cause this effect will depend on the strength of the relaxation processes. They considered a three-level scheme $E_3 > E_2 > E_1$, in which intense radiation was applied at $h\nu_{13} = E_3 - E_1$ so that $n_3' = n_1'$. If the spacing of the energy levels is correct it is possible to obtain the condition $n_2 > n_1'$. This inverted population could then be used to cause oscillations or amplification. No device working on this system of levels appears to have been made, but such a device would operate at fixed frequencies.

Bloembergen (1956) independently made a similar suggestion based on a three-level scheme, but this time the levels concerned were those of a paramagnetic ion contained in a crystal subjected to a magnetic field. This scheme had the tremendous advantage that the levels could be tuned by the adjustment of the magnetic field. In addition a considerably higher concentration of active ions could be produced compared with the density in a gas, thus resulting in a greater emission of energy. Bloembergen considered in

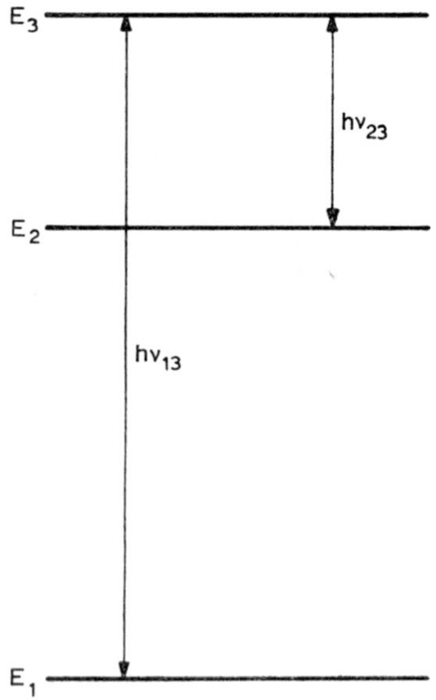

FIG. 1. Energy level scheme.

detail the conditions required by writing equations representing transition rates between the levels caused by the radiation and by relaxation processes. Solution of these 'rate equations' yielded an expression for the resultant population inversion in terms of the frequencies used and the rates of the relaxation processes.

The rate equations can be illustrated with reference to Fig. 1 which shows three energy levels, E_1, E_2 and E_3, subjected to radio frequency fields at the frequencies $\nu_{13} = (E_3 - E_1)/h$, termed the pump frequency, and $\nu_{23} = (E_3 - E_2)/h$, the signal frequency, which cause stimulated transitions at rates W_{13} and W_{23} respectively. In addition to these transitions, there will be other thermalising processes between the levels which can be represented by transition rates w_{12}, w_{21}, etc., for upward and downward transitions. These thermalising rates represent the interaction of the lattice vibrations with the paramagnetic system. The spontaneous emission taking place at microwave frequencies can be neglected as shown by equation (1.7).

The rates of change of the level populations may then be written:

$$\frac{dn_1}{dt} = -n_1(w_{12}+w_{13})+n_2w_{21}+n_3w_{31}+(n_3-n_1)W_{31} \tag{1.10}$$

$$\frac{dn_2}{dt} = -n_2(w_{23}+w_{21})+n_3w_{32}+n_1w_{12}+(n_3-n_2)W_{23} \tag{1.11}$$

$$\frac{dn_3}{dt} = -n_3(w_{31}+w_{32})+n_2w_{23}+n_1w_{13}-(n_3-n_1)W_{13} +(n_2-n_3)W_{23} \tag{1.12}$$

in which the first three terms on the right represent the effects of thermalising processes and the remaining terms represent the influence of the applied fields.

In addition to these three equations the equality

$$n_1+n_2+n_3 = N \tag{1.13}$$

where N is the total number of active paramagnetic ions, must also hold, though this is not an independent equation, but is implicit in equations (1.10) to (1.12).

In the absence of the radio frequency fields, any pair of levels would revert to a Boltzmann distribution, in which case the upward and downward thermal processes balance giving:

$$n_{1_0} w_{12} = n_{2_0} w_{21}, \text{ etc.,} \tag{1.14}$$

n_{1_0}, n_{2_0} being the equilibrium values of the populations. Thus:

$$\frac{w_{12}}{w_{21}} = \frac{n_{2_0}}{n_{1_0}} = \exp-\left(\frac{h\nu_{12}}{kT}\right) \tag{1.15}$$

Similar relations hold for the other thermal rates:

$$\frac{w_{13}}{w_{31}} = \exp-\left(\frac{h\nu_{13}}{kT}\right) \tag{1.16}$$

$$\frac{w_{23}}{w_{32}} = \exp-\left(\frac{h\nu_{23}}{kT}\right) \tag{1.17}$$

If the pump power applied at ν_{13} is made very large so that $W_{13} \gg W_{23}, w_{12}, w_{21}, w_{23}$, etc., then at dynamic equilibrium when

$$\frac{dn_1}{dt}, \frac{dn_2}{dt}, \frac{dn_3}{dt} = 0$$

equation (1.10) shows that $n_1 = n_3$ (i.e. the transition $1 \rightarrow 3$ is saturated). Making this substitution in (1.11) gives:

$$0 = -n_2(w_{23}+w_{21}+W_{23})+n_3(w_{32}+w_{12}+W_{23}) \tag{1.18}$$

from which:

$$\frac{n_1}{n_2} = \frac{n_3}{n_2} = \left(\frac{w_{23}+w_{21}+W_{23}}{w_{32}+w_{12}+W_{23}}\right) \tag{1.19}$$

Then:

$$\frac{n_3-n_2}{N} = \frac{n_1-n_2}{N} = \frac{(n_3/n_2)-1}{(2n_3/n_2)+1}$$

$$= \frac{(w_{23}-w_{32})+(w_{21}-w_{12})}{w_{32}+2w_{23}+w_{12}+2w_{21}+3W_{23}} \tag{1.20}$$

since $N = n_1+n_2+n_3 = n_2+2n_3$.

Using equations (1.15), (1.16) and (1.17) this may be written:

$$\frac{n_3-n_2}{N} = \frac{n_1-n_2}{N}$$

$$= \frac{w_{23}(1-\exp(h\nu_{23}/kT))-w_{12}(1-\exp(h\nu_{12}/kT))}{w_{23}(2+\exp(h\nu_{23}/kT))+w_{12}[2\exp(h\nu_{12}/kT)+1]+3W_{23}} \tag{1.21}$$

This equation may be simplified by making the linear approximation:

$$\exp\frac{h\nu_{ij}}{kT} \simeq 1+\frac{h\nu_{ij}}{kT} \quad \text{valid for} \quad \frac{h\nu_{ij}}{kT} \ll 1$$

This is usually justified for microwave frequencies in the region of 10 GHz where $h\nu/k \simeq 0{\cdot}45$°K (thus even at temperatures as low as 2·0°K. $h\nu/kT = 0{\cdot}23$), but care must be exercised in applying it when higher frequencies are involved.

In this approximation equation (1.21) becomes:

$$\frac{n_3-n_2}{N} = \frac{n_1-n_2}{N} = \frac{h(w_{12}\nu_{12}-w_{23}\nu_{23})}{3kT(w_{23}+w_{12}+W_{23})} \tag{1.22}$$

and it can be seen that if $w_{12}\nu_{12} > w_{23}\nu_{23}$ the condition for emission at ν_{23} ($n_3 > n_2$) can be obtained. This reversal of population or inversion, as it came to be called, is often expressed as a ratio of the population differences obtained with and without the pump applied:

$$\text{Inversion } (I) = \frac{(n_3-n_2)\text{ pump applied}}{(n_3-n_2)\text{ no pump}} \tag{1.23}$$

Making the assumption that $w_{12} \simeq w_{23} \gg W_{23}$ and remembering that at thermal equilibrium:

$$n_3-n_2 \simeq -\frac{Nh\nu_{23}}{3kT} \tag{1.24}$$

the inversion in this case becomes:

$$I = \frac{\nu_{13}}{2\nu_{23}} - 1 \tag{1.25}$$

showing that for large inversions the pumping frequency ν_{13} must be much greater than twice the signal frequency. Figure 2 shows the populations before and after the application of the pump and illustrates qualitatively how inversion may be obtained between levels 2 and 3. However, it should not be taken too literally as it assumes n_2 to remain constant, which is not strictly valid.

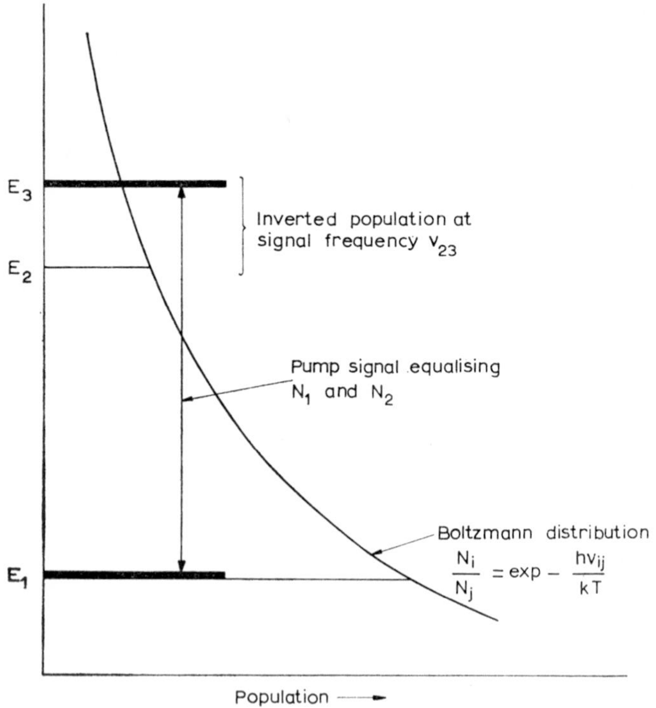

FIG. 2. Inversion at ν_{23} produced by saturation of 1–3 transitions by pumping.

The power output from the material is obtained by combining equations (1.9) rewritten for levels 2 and 3 and (1.22) giving:

$$P_{out} = \frac{Nh^2\nu_{23}}{3kT}\frac{(w_{12}\nu_{12}-w_{23}\nu_{23})W_{23}}{[w_{23}+w_{12}+W_{23}]} \tag{1.26}$$

Since $W_{23} = B_{23}\,\rho(\nu_{23})$ the power out will vary as the stimulating energy density varies thus providing the necessary conditions for amplification. Note that it is important that $W_{23} \ll w_{23}, w_{12}$, etc., otherwise the population difference n_3-n_2 can be reduced by the saturating effects of the signal, thus the device operates most efficiently at low input power. Note also that for maximum power a high concentration N of active atoms is required, together with a low ambient temperature T.

Comparatively little detail was known about the behaviour of paramagnetic systems under such conditions, but using data available for the nickel ion Ni^{2+} in a host lattice of zinc fluosilicate and for Gd^{3+} in lanthanum ethyl sulphate, Bloembergen showed that sufficient power should be emitted to overcome circuit losses and thus enable the device to operate as an oscillator or amplifier in the microwave region. He also pointed out that in the latter the low temperature of operation and the low effective temperature of the inverted spin population would make the device an almost ideal low noise amplifier.

Immediately a number of laboratories began experimenting on a variety of paramagnetic materials and the first realisation as an oscillator was reported by Scovil, Feher and Seidel (1957), who saw the onset of oscillations caused by emission from the inverted levels of Gd^{3+} in a host lattice of lanthanum ethyl sulphate. It is interesting to note that in this material for a signal at ν_{12}, the inversion requirement $\nu_{23}w_{23} \gg \nu_{12}w_{12}$ was obtained by altering the ratio of w_{12} to w_{23} by the introduction of Ce^{3+} into the crystal. This technique of cross doping was necessary because the energy levels used were such that $\nu_{12} \simeq \nu_{23}$ for the orientation and field chosen. This cross-relaxation process is further discussed in Chapter 2.

The first amplifier was reported by McWhorter and Meyer (1958) using Cr^{3+} in potassium cobalticyanide and, later the same year, Makhov *et al.* (1958) operated a maser using ruby (Cr^{3+} in Al_2O_3) as the active material. This was an important development as ruby, being a hard, durable material, was ideally suited for the practical realisation of a microwave device and indeed has been the most frequently used maser material to date.

The first solid state masers were usually in the form of a microwave cavity which contained the paramagnetic salt. A schematic diagram of such a cavity maser in an experimental arrangement is shown in Fig. 3. The paramagnetic sample is contained in the

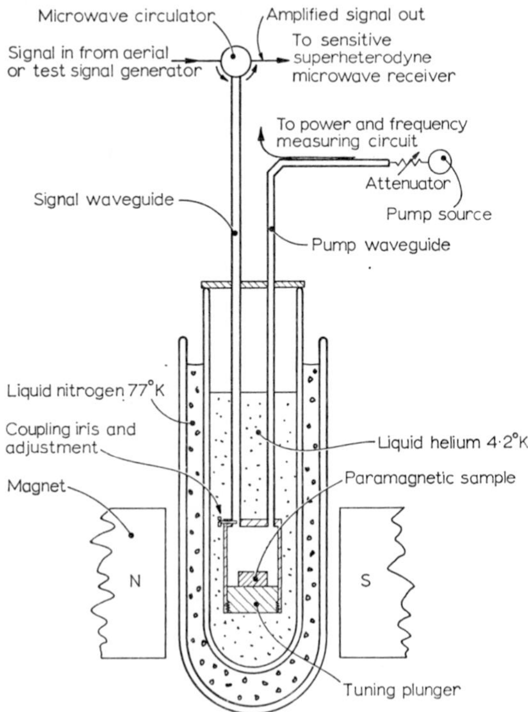

FIG. 3. Schematic diagram of simple cavity maser system.

microwave cavity which can be tuned to the signal frequency by means of a movable plunger at the base. The signals to be amplified are fed to the cavity via an iris which in an experimental set-up is usually adjustable. The input and output signals are separated by the microwave circulator at the top of the waveguide. The pump energy enters via a second waveguide and the cavity may or may not be made to resonate at the pump frequency simultaneously with the signal frequency, depending on the power available from the pump source. The whole system is immersed in liquid helium contained in the double dewar arrangement shown. In an experimental device the magnetic field is applied usually by means of an electromagnet which, for ease of examination of various modes of operation, is arranged to rotate about a vertical axis through the dewar system. In a practical device the magnetic field is usually produced by a permanent magnet, or in some cases by means of a superconducting magnet built around the maser and immersed in the liquid helium. These cavity masers suffered from the disadvantages that they could easily break into oscillation if the cavity coupling conditions varied and, being regenerative devices, had a fixed gain–bandwidth product. Later advances stemmed from the replacement of the cavity with a slow wave structure to propagate the signal frequency. Such a slow wave structure is illustrated in Fig. 4. The signal is propagated along an array of parallel conductors ("fingers") contained within a waveguide in which the pump energy is also propagated. The geometry of the conductors is chosen so that the signal velocity is reduced to the order of one-hundredth of the free space velocity. The microwave fields at the signal frequency are distributed around the conductors, and the paramagnetic material is placed against one side of the fingers as shown.

As the signal enters and leaves by separate leads the problem of separating the input and output is removed. In addition, any tendency for the device to oscillate due to end-to-end reflections can be suppressed by the incorporation of ferrite elements on the side of the fingers away from the paramagnetic sample. By arranging the operating magnetic field along the direction of the fingers

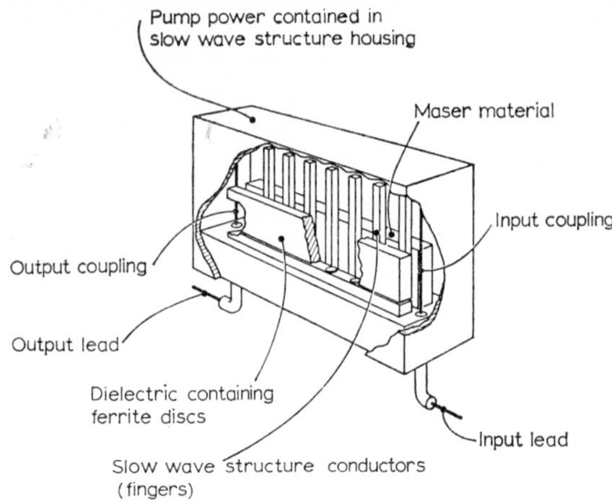

FIG. 4. Basic construction of a travelling wave maser.

and positioning the ferrite material correctly these can be made to act as isolators by producing loss for waves travelling back from the output towards the input. By making this loss greater than the gain of the device oscillations can be suppressed. As the maser may be up to 15 cm long the provision of the magnetic field may be difficult, and specially shaped permanent magnets, or superconducting magnets containing special shields, to keep the field uniform over the volume of the maser material are used. These travelling wave masers (T.W.M.) were employed in a number of successful applications. Their construction and operation is more fully described in Chapter 3.

In Part II the first five papers have been chosen to illustrate some of the developments prior to Bloembergen's 1956 proposal. They extend over the period 1950–6 starting with Purcell and Pound's observation of inversion in a nuclear spin system. The second and third papers show Weber's and Basov and Prokhorov's suggestions for the production of amplifying systems using inverted populations. The fourth paper shows the first results reported by Gordon, Zeiger and Townes on the ammonia maser and finally Bloembergen's paper is presented.

CHAPTER 2

Maser Materials

2.1. Paramagnetic Resonance*

Following the historical introduction in Chapter 1, we now concentrate attention on the materials used in making solid state masers, i.e. paramagnetic ions dissolved in a dielectric crystal lattice. In particular, we shall be concerned with the interaction of these ions with the host crystal and with an applied magnetic field.

A paramagnetic ion possesses a permanent magnetic moment $\boldsymbol{\mu}$ due to the presence of one or more unpaired electrons. A completely filled electron shell has zero net moment as all the individual electron moments are "paired", i.e. they cancel one another. It is only when partially filled shells occur, as happens in transition group ions, that a permanent magnetic moment results.

If such an ion is placed in a uniform magnetic field $\boldsymbol{H}$ there is an energy of interaction given by $\boldsymbol{\mu} \cdot \boldsymbol{H}$ just as for the case of a bar magnet. However, on the atomic scale we must take account of the fact that energy is quantised and only a limited number of orientations of the moment are allowed. In the simplest case $\boldsymbol{\mu}$ may be aligned either parallel or anti-parallel to $\boldsymbol{H}$ with corresponding energies $\pm \boldsymbol{\mu H}$. We can represent this by the energy level diagram of Fig. 5.

It is possible to induce transitions between these two allowed states of the system (i.e. to flip the magnetic moment over) by

* For more detailed accounts of the subject of paramagnetic resonance in crystals, the reader is referred to the books by Low (1960), Pake (1962) and Orton (1968). For experimental techniques see also the books by Ingram (1967), Assenheim (1966) and Poole (1967).

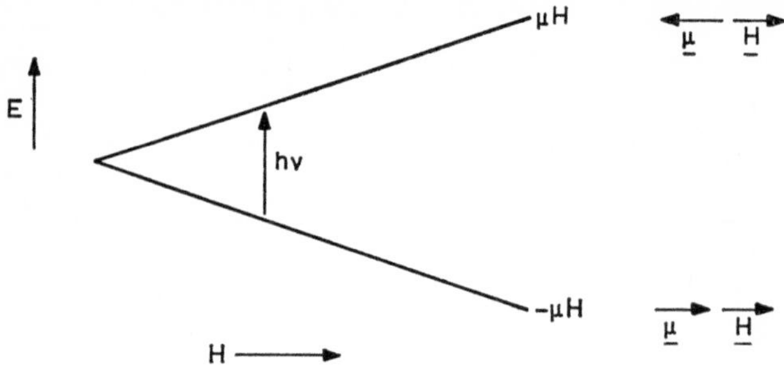

FIG. 5. Energy level diagram for a magnetic moment μ in a uniform field H when only two orientations are allowed quantum mechanically.

applying radiation of a suitable frequency such that the energy of a single photon is equal to the difference in energy of the two states. This defines a resonance condition:

$$h\nu = 2\mu H_0 \tag{2.1}$$

and illustrates the principle of "paramagnetic resonance", energy being exchanged between the radiation field and the paramagnetic ion. For the simple case where μ is the moment associated with the spin of a single electron $\mu \equiv \beta$, where β is the Bohr magneton having the value $9{\cdot}273 \times 10^{-21}$ ergs/gauss. As the electron also possesses a component of magnetic moment due to its orbital motion in the ion equation (2.1) must be modified and it is usual to write:

$$h\nu = g\beta H_0 \tag{2.2}$$

where g is known as the spectroscopic splitting factor having a value 2 for pure spin magnetism and lying within the range 1 to 10 for most known cases.

Assuming $g = 2$ equation (2.2) gives the order of magnitude of frequency involved. Thus if $H_0 = 1$ kG, $\nu \approx 3$ GHz and we see that energy level "splittings" of the desired magnitude (i.e. in the microwave region) are obtained with easily attainable magnetic fields.

2.2. Paramagnetic Ions in Crystals

Having established the principle of paramagnetic resonance we must now consider the properties of the magnetic ions in more detail. Practical solid state masers operate at low temperatures (usually 4°K or below) where the only energy levels populated are those within a few wave numbers of the lowest. We first consider the ground state levels of a free paramagnetic ion and then the way in which these are modified when the ion is contained in a crystal·

In the following we shall take into account only the unpaired electrons as these are responsible for the magnetic properties, but we must now include the possibility that there are several of them. In this case the usual procedure is to use the vector model of the atom to determine the total angular momentum and magnetic moment of the ion. Each electron is described by orbital and spin angular momentum quantum numbers l and s, the corresponding momenta being $l(h/2\pi)$ and $s(h/2\pi)$. Assuming Russell–Saunders coupling, the orbital momenta add to give a total orbital momentum $L(h/2\pi)$ and the spins are similarly coupled to give momentum $S(h/2\pi)$. These in turn are coupled by spin–orbit interaction to give a total angular momentum for the ion $J(h/2\pi)$.

In general there are several possible ways of combining momenta, to give a series of allowed energy states separated in energy by $\sim 10^4$–10^5 cm^{-1} (corresponding to different values of L, S and J), but we are only interested in the combination giving lowest energy. This may be found from Hund's rules which state:

1. The total spin angular momentum should be maximum.
2. The total orbital momentum should also be maximum consistent with the Pauli exclusion principle that no two electrons should have all their quantum numbers identical.

This is most easily explained by an example. Consider the trivalent chromium ion Cr^{3+} which has the electron configuration $(1s^2\ 2s^2\ 2p^6\ 3s^2\ 3p^6)\ 3d^3$ (in future we shall neglect all closed shells and refer to this simply as $3d^3$). The unpaired electrons, having $l = 2$ and $s = \frac{1}{2}$ couple to give minimum energy as shown

TABLE 1

Electron	s	m_s	l	m_l	S	L
1	$\frac{1}{2}$	$\frac{1}{2}$	2	2		
2	$\frac{1}{2}$	$\frac{1}{2}$	2	1	3/2	3
3	$\frac{1}{2}$	$\frac{1}{2}$	2	0		

in Table 1 giving a free ion ground state $^4F_{3/2}$ (i.e. $S = 3/2$, $L = 3$, $J = 3/2$) the convention used being $^{(2S+1)}L_J$. A fourth $3d$ electron (as in Cr^{2+}) would have $m_s = \frac{1}{2}$, $m_l = -1$ giving a ground state $^5D_4$ and so on up to $3d^{10}$ where the shell is full and $S = L = 0$.

The ground state of the free ion is $(2J+1)$ fold degenerate, but this degeneracy is lifted by a magnetic field since, associated with the angular momentum $J(h/2\pi)$ is a magnetic moment $\mu = g_L\beta J$ (where g_L is the Landé g-factor, see Hindmarsh (1967)) which may take up $(2J+1)$ orientations with respect to $\boldsymbol{H}$. Transitions may be induced between the levels corresponding to these orientations in the same way as described above, but we shall not consider this situation further as important modifications are introduced when the ion is situated in a crystal.

The interaction between the magnetic ion and its neighbours in the crystal is usually treated by the method of crystal field theory. In this approximation the neighbouring ions may be regarded as point electric charges located at appropriate lattice sites and giving rise to an electric field in the vicinity of the magnetic ion. The unpaired electrons on the magnetic ion are thus moving in a spatially varying field and their energies are modified accordingly. The important point is that different orbits (characterised by different values of m_l) have markedly different shapes and are therefore affected to different extents. Whereas in the free ion all electrons with the same value of l had the same energy this is no longer true when the ion is in a crystal—the crystal field raises the orbital degeneracy giving rise to "crystal field splittings" of the orbital levels. The precise details will obviously depend on the symmetry of the lattice site containing the magnetic ion.

Let us return to our example. As before, the three unpaired electrons couple to give orbital and spin angular momenta $L(h/2\pi)$ and $S(h/2\pi)$ but the interaction of the orbital motion with the crystal field is much larger than that between spins and orbits,

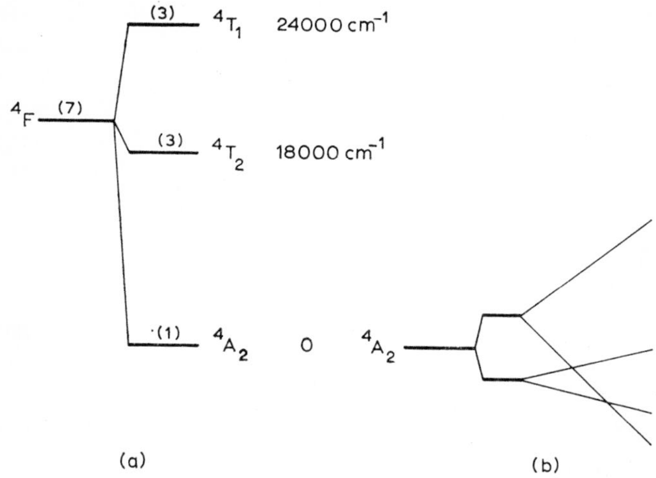

FIG. 6. Energy level splittings for the Cr^{3+} ion in ruby. (*a*) Splitting of the orbital levels in a crystal field of octahedral symmetry. (*b*) Further splitting of the $^4A_2$ levels due to the effect of a trigonal distortion and applied magnetic field.

so spin–orbit coupling is "broken down" and it is no longer possible to speak of a total angular momentum. This is often referred to by saying that J is no longer a "good quantum number". The crystal field raises the $(2L+1)$ fold orbital degeneracy $(2L+1 = 7$ in this case) in the manner illustrated in Fig. 6(*a*) where we have assumed a lattice site having 6-fold (octahedral) coordination giving rise to two triplet levels $^4T_1$ and $^4T_2$ and a singlet $^4A_2$. (The labelling of the levels derives from group theory (Tinkham, 1964),

T implying a triplet orbital state and A a singlet. The superscript is the spin degeneracy $(2S+1)$ as before.

The splittings produced are of order 10^4 cm^{-1} so, as we are concerned with the ground state we need consider only the $^4A_2$ levels which are four-fold degenerate due to the spin. In crystal fields of accurately cubic symmetry (i.e. regular octahedral surroundings) this degeneracy is only raised by a magnetic field, but in fields of lower symmetry the ground state consists of two doublets separated by energies of the order of 3–50 GHz. This final splitting, known as the "zero field splitting" (i.e. zero magnetic field), arises by way of spin–orbit coupling as an electric field does not interact directly with the electron spin. It is for this reason that the splitting is very small. The application of a magnetic field further splits the doublets leaving four singlet levels (see Fig. 6(*b*)) separated by irregular intervals but all lying within a range of microwave energies. This grouping is of precisely the kind required for a three-level maser.

The importance of the zero-field splitting may be appreciated by considering the implication of equation (1.25). Since it is advantageous to use a pump frequency ν_{13} considerably greater than twice the signal frequency ν_{12}, it is convenient to employ a material where the zero field splitting is a few times the signal frequency.

This situation contrasts with that holding for most rare earth ions (the second important transition group). For example, let us consider the Nd^{3+} ion which has three unpaired $4f$ electrons giving rise to a free-ion ground state $^4I_{9/2}$. The $4f$ shell is not, however, the outermost occupied shell as is true of the $3d$ shell in the iron group. The $5s$ and $5p$ shells are filled and these outer electrons form an electrostatic screen which shields the $4f$ electrons from the influence of the crystal field. Thus, spin–orbit coupling is much stronger than the crystal field interaction and to a first order approximation the energy levels of the Nd^{3+} ion are similar to those of the free ion. The lowest group of levels is designated $^4I_{9/2}$ while the first excited group corresponding to a different J value is $^4I_{11/2}$ lying roughly 1000 cm^{-1} higher. The effect of the crystal field is to raise the degeneracy of these multiplets. Thus, the $J=9/2$ multiplet is split by fields of axial or lower symmetry into five doublets (known as Kramers doublets) with separations of order 100 cm^{-1} and the only level occupied at low temperature is the lowest of these. This isolated doublet is unsuitable for three-level maser pumping, at least so long as there are no sources of power in the sub-millimeter wavelength region. The situation is illustrated in Fig. 7.

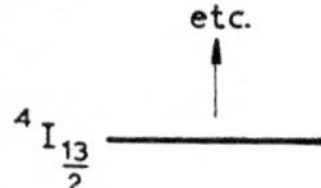

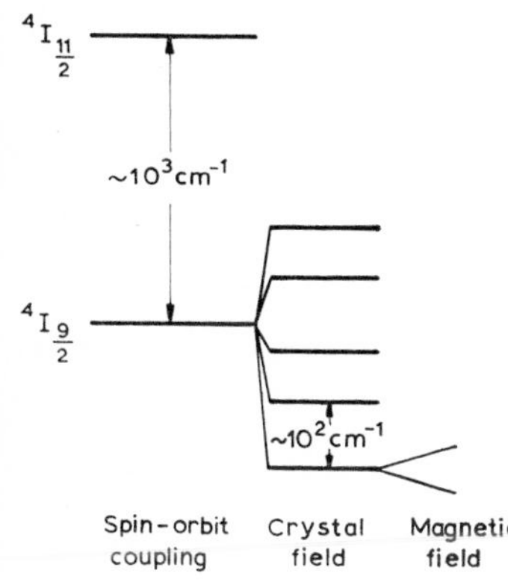

FIG. 7. Splitting of the Nd^{3+} levels due to spin–orbit coupling, crystalline field and applied magnetic field. The ground state of the ion in a crystal is an isolated Kramers doublet.

2.3. Choice of Maser Material

We see from the above outline that the nature of the lowest energy levels depends both on the paramagnetic ion in question and the type of crystal in which it is situated. It is important to bear this in mind in any discussion of maser materials, but we shall make a rather arbitrary division into the two components in the following section.

2.3.1. *The active ion*

The first important property of the magnetic ion is that it should possess a ground state (when in the crystal) with three or more levels. The details of these ground state levels are, of course, dependent on the host crystal but, as we saw above, the majority of rare earth ions is ruled out.

Another requirement is that the ionic energy levels should not be split by interaction with a nuclear magnetic moment. Thus, if the nucleus of the ion in question possesses a magnetic moment each line of the magnetic resonance spectrum is split into $(2I+1)$ components (referred to as hyperfine structure) where I is the appropriate nuclear spin. Only one of these resonance lines may be used in the maser so the effect of hyperfine structure is to reduce the effective number of active ions by the factor $(2I+1)$. A glance at any table of the magnetic properties of nuclei (e.g. Assenheim, 1966) serves to show that this considerably reduces the number of potentially useful ions. For example, the V^{2+} ion has the same electronic configuration as Cr^{3+} but the ^{51}V nucleus, which is almost 100% abundant in naturally occurring vanadium, has a spin $I = 7/2$ and a large magnetic moment.

These two requirements alone reduce the likely candidates for acceptance as maser ions to a very small number indeed, namely Cr^{3+} $(3d^3)$, Fe^{3+} $(3d^5)$, Ni^{2+} $(3d^8)$ and Gd^{3+} $(4f^7)$. This last ion, though a rare earth, qualifies as its free ion ground state is an S-state $(^8S_{7/2})$ which is not affected by a crystal field except to a very small extent by way of a "high order" spin–orbit interaction (as the orbital angular momentum is zero there can be no direct interaction with an electric field). All these ions have an orbital singlet level lowest in a crystal field of octahedral symmetry and a maser may operate using the spin levels where $S = 3/2$, $5/2$, 1 and $7/2$ respectively.

2.3.2. *The host crystal*

The first restriction on the host crystal is that it must be diamagnetic so as not to interfere with the magnetic properties of the maser ion. It must possess good mechanical properties (i.e. be capable of machining), be stable in air and able to withstand the thermal shock of cycling between room and liquid helium temperatures. Another important requirement is the availability of large single crystals (practical maser crystals may be anything up to 12 cm in length by 1 cm^2 in cross-section) containing a controlled amount of maser impurity ions. This implies that it must be possible to grow crystals artificially.

In general the only suitable materials are crystals of gem-stone type which are hard, stable and resilient and this imposes a severe restriction in that crystal-growing techniques have been developed for only a limited number. For example, alumina (Al_2O_3) and rutile (TiO_2) may be grown by the Verneuil method (melting of powder dropped through an oxyhydrogen flame) and calcium and zinc tungstates ($CaWO_4$ and $ZnWO_4$) and yttrium aluminium garnet ($Y_3Al_5O_{12}$) by the Czochralski method (pulling from the melt). See Brice (1965) for details of crystal growth.

When one considers the marriage of the impurity ion and host crystal to produce the complete maser material one or two other points must be noted. In particular, there must be a suitable site available in the crystal and this implies the presence of a metal ion which can readily be replaced by the maser ion. It is a general rule that these ions should be similar in size. Thus, it is possible to substitute the Cr^{3+} ion (ionic radius 0·65 Å) for the Al^{3+} ion in Al_2O_3 (radius 0·55 Å) but almost impossible to incorporate the larger Gd^{3+} ion (radius 1·1 Å). The latter is much more readily substituted for Ca^{2+} or Y^{3+} which match it fairly closely in size.

The site symmetry has already been mentioned. This determines the detailed behaviour of the ionic ground state and for the ions Cr^{3+} and Ni^{2+} should be approximately octahedral in order that the desired orbital level should be lowest. It is also important that the octahedron be somewhat distorted to produce a zero-field splitting of the spin levels, but this condition is satisfied by a large number of materials.

One final limitation is that there should be at most two and preferably only one impurity site in the unit cell of the host crystal.

In many crystals, such as yttrium aluminium garnet, the impurity ion may enter several magnetically inequivalent sites which may differ in having distinct symmetries (in Y.A.G. there are both octahedral and tetrahedral Al sites) or simply in their orientation, with respect to the crystal axes. In the latter case they become inequivalent only when a magnetic field is applied. The existence of these alternative sites reduces the effective number of magnetic ions taking part in maser action which is obviously undesirable.

2.4. Properties of Maser Energy Levels

2.4.1. *Energies and transition probabilities*

Having enumerated some of the practical points involved in selecting a maser material we must now consider in more detail the properties of the energy levels involved in maser action. In particular, it is important to have some way of calculating the relative positions of these levels and the strengths of transitions induced between them by a microwave field. The formal treatment of the problem employs the spin Hamiltonian method which we outline later. For the present we shall be content with a qualitative description.

As already indicated, the ions suitable for use in masers possess a group of energy levels closely spaced among themselves, but separated from the next higher group by several hundred cm^{-1} (or more). It is convenient then to define an "effective spin" S such that the multiplicity of the ground state is equal to $(2S+1)$. In many cases S is identical with the actual spin of the free ion but this is not necessarily so. For example, many rare earth ions with an odd number of $4f$ electrons have ground states (in the crystal) consisting of a single Kramers doublet which must be described by an effective spin $S = \frac{1}{2}$ though the free ion may have spin 3/2, 5/2, etc.

We may also define a magnetic quantum number M in the usual way such that M takes values S, $(S-1)$, . . ., $-S$ and each of the $(2S+1)$ levels may be characterised by a particular value of M.

The significance of M is that, in the absence of crystal field effects, the energies of the levels in a magnetic field H are given by:

$$E_M = \tfrac{1}{2}Mg\beta H \qquad (2.3)$$

Transitions between the levels may be induced by a microwave field provided the selection rule $\Delta M = \pm 1$ is satisfied, the resonance condition being given by equation (2.2). It is apparent that all $2S$ transitions will then occur at the same magnetic field (assuming a fixed microwave frequency) and will be indistinguishable.

The situation may be quite different when the effects of the crystal field are included. As already mentioned in Section 2.2 the ground state degeneracy may be partially lifted even in the absence of a magnetic field and this crystal field splitting gives rise to two important effects. On application of a steady magnetic field the levels again diverge as suggested by equation (2.3) but the allowed transitions ($\Delta M = \pm 1$) no longer occur together; the paramagnetic resonance spectrum exhibits "fine structure". This is illustrated for the case of Cr^{3+} ($S = 3/2$) in an axial crystal field

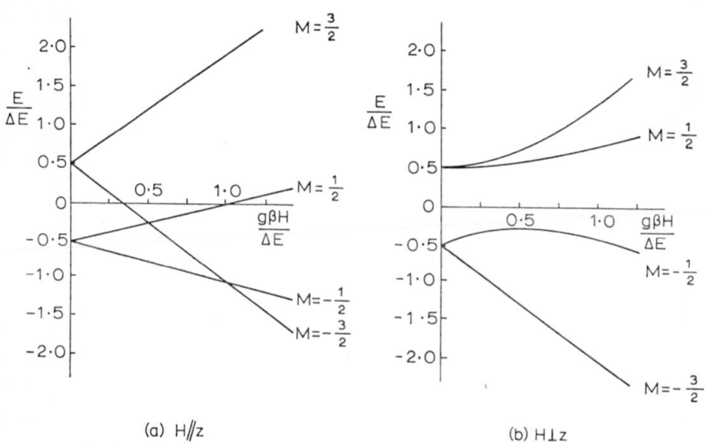

FIG. 8. Normalised plots of the Cr^{3+} ground state levels in axially symmetric crystal symmetry (*a*) with magnetic field applied parallel to the symmetry axis, (*b*) with H perpendicular to the symmetry axis. ΔE is the zero-field splitting.

in Fig. 8(*a*) where the magnetic field is applied parallel to the symmetry axis of the crystal field (usually taken as the Z-axis). The second important difference is illustrated by Fig. 8(*b*) where the magnetic field is applied perpendicular to Z, the energies no longer being linear functions of H. This is due to the opposing tendencies of the crystal and magnetic fields, the crystal field trying to align the spin magnetic moment parallel to Z while the magnetic field tries to align it perpendicular to Z. At large values of H the magnetic field dominates and the levels again become straight, but the intermediate region where the magnetic and crystal field interactions are comparable (i.e. $g\beta H \approx \Delta E$) is of special interest for maser operation for in this region the selection rule $\Delta M = \pm 1$ is broken down. This arises because the crystal field mixes the wave functions of the various energy states so that it is no longer correct to designate the upper level (for example) by $M = 3/2$, it now possesses some character associated with the other values of M. Formally, we may write the wave function of each level as:

$$|\,\psi_i\rangle = a_i\,|\,M = \tfrac{3}{2}\rangle + b_i\,|\,M = \tfrac{1}{2}\rangle + c_i\,|\,M = -\tfrac{1}{2}\rangle + d_i\,|\,M = -\tfrac{3}{2}\rangle \qquad (2.4)$$

and it can be seen that as the wave function of each level contains admixtures of all M values, the $\Delta M = \pm 1$ rule implies that transitions are now possible between any pair of levels. This is important for three-level maser operation as it is essential for the pump transition $(1 \rightarrow 3)$ to be allowed.

These considerations apply to any orientation of applied field except the limiting case $H \parallel Z$ and the details of state mixing (i.e. the values of a_i, b_i, c_i and d_i in equation (2.4)) will depend on the magnitude and orientation of H. Thus we see that this system provides admirable flexibility as it offers the possibility of controlling both the relative positions of the levels and the transition rates between them by varying the applied magnetic field.

2.4.2. *Line shapes*

So far in our discussion we have spoken as though resonance between magnetic ion and microwave field occurs at a precise field H_0 and is infinitely sharp. This is not, of course, true in practice and resonance occurs over a small range of fields characterised by a line shape and line width ΔH. There are various sources of line broadening in paramagnetic resonance, usually divided into the two categories of homogeneous and inhomogeneous broadening.

An example of homogeneous broadening is provided by dipolar interaction between similar magnetic ions in the crystal. In the presence of an applied field H the magnetic moment vectors of the individual ions precess about H at the Larmor precession fre-

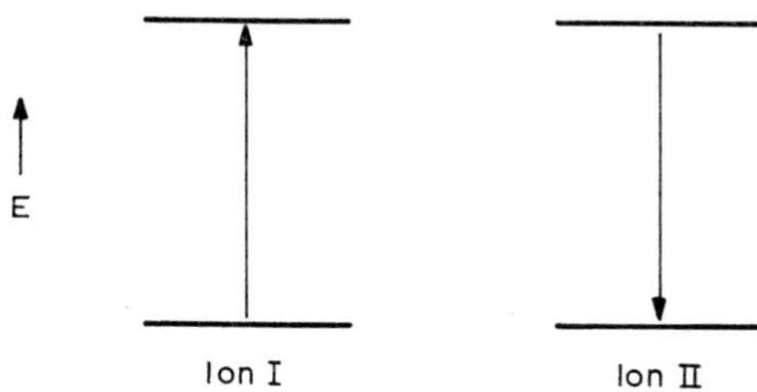

FIG. 9. Illustration of "flip-flop" process involved in dipolar interaction between like ions. There is no net change in level population.

quency (Pake, 1962, chap. 2) and if we consider any one specific ion it will experience a field due to all the others which is varying at this frequency due to the precession. The effect is similar to that of an applied microwave field in inducing a transition between two states of the ion in question. Energy is conserved within the system as a whole—for every ion making an upward transition a second will reverse the process so there is no net change in the populations of the levels (see Fig. 9). However, the effect is to limit the average time an ion remains in any particular level and results in "lifetime broadening" of the levels according to the Uncertainty Principle:

$$\Delta E \, \Delta t \sim h/2\pi \tag{2.5}$$

It is usual to define a characteristic time T_2 for dipolar broadening which is related to the linewidth $\Delta\nu$ according to:

$$\Delta\nu = 1/\pi T_2$$

where $\Delta\nu$ is the full width of the resonance line at half height.

Lifetime broadening is always characterised by a Lorentzian line shape which may be described by the function:

$$g(\nu) = \frac{2T_2}{1+4\pi^2(\nu-\nu_0)^2T_2^2} \tag{2.6}$$

where ν_0 is the exact resonance frequency as given by equation (2.2) and $g(\nu)$ is normalised so that $\int_0^\infty g(\nu)d\nu = 1$. In the case of a dilute solution of paramagnetic ions in a diamagnetic crystal ($\sim$ 0·05%) a typical value of T_2 would be 10^{-8} sec corresponding to a line width $\Delta\nu \approx 30$ MHz. This may be converted to a magnetic field width ΔH by the relation:

$$\Delta H = \frac{\partial H}{\partial \nu}\,\Delta\nu \tag{2.7}$$

which, for the simple case where equation (2.2) holds, gives a value of $\Delta H \approx 10$ gauss (assuming $g = 2$). However, as discussed above, the relation between H and ν is not necessarily linear and $\partial H/\partial \nu$ must be obtained graphically (see, for example, Fig. 8(*b*)). The frequency width $\Delta\nu$ is the more fundamental parameter.

Another source of lifetime broadening is the interaction of the paramagnetic ions (spins) with lattice vibrations—spin–lattice relaxation—which we shall discuss in more detail in the next section. This is usually described in terms of a second characteristic time T_1.

As an example of inhomogeneous broadening we may cite the effect of imperfections in the crystal which give rise to small variations in the crystal field from one paramagnetic ion to another. Thus, the zero-field splitting will be "smeared out" to a certain extent and a glance at Fig. 8 shows that the energy levels in an applied magnetic field will be broadened in consequence. Even though the natural line width for transitions of any particular ion may be sharp the overall width observed will be that of the envelope made up of a series of displaced lines as shown in Fig. 10. This type of inhomogeneous broadening is usually characterised by a random distribution of imperfections giving rise to a Gaussian line shape.

Other sources of inhomogeneous broadening are dipolar interaction with "unlike" magnetic dipoles (such as the nuclear moments of neighbouring ions) and inhomogeneities in the applied magnetic field.

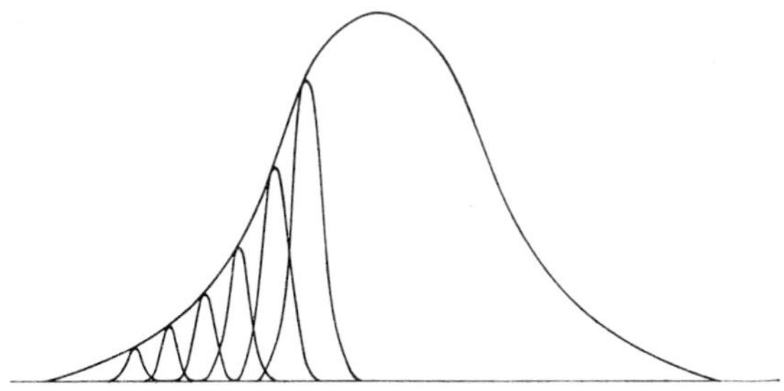

FIG. 10. Inhomogeneously broadened line, the line shape being the envelope of a large number of slightly displaced sharp components.

In practice, the actual line width may arise as a sum of contributions from several sources and the shape is often intermediate between Lorentzian and Gaussian. It will, in general, depend on the concentration of magnetic ions (which affects the strength of dipole–dipole and exchange interactions) and on temperature, as spin–lattice interaction depends strongly on temperature.

2.5. Spin–lattice Relaxation

So far we have considered the paramagnetic ions (often referred to as the "spin-system") as interacting with a static crystal lattice. To make our picture more realistic we must modify this assumption. The ions of the host crystal vibrate relative to one another with energies characteristic of the lattice and its temperature. The energy levels of a paramagnetic ion in the crystal depend on the electric field produced by its neighbours, and thus on their relative positions. Thermal vibrations modulate these positions, and hence the magnetic energy levels, in a manner rather similar to that produced by applying an oscillating magnetic field. It is therefore possible for lattice vibrations to induce transitions between the paramagnetic levels and energy is thus exchanged between the spin system and the lattice. This interaction tends to maintain the spins in thermal equilibrium with their surroundings and, in the absence of any perturbation such as an oscillating magnetic field close to the resonance frequency, results in the populations of the paramagnetic energy levels being governed by the well known Boltzmann distribution. The population of the ith level is proportional to $\exp\{-(E_i/kT)\}$.

To illustrate the importance of spin–lattice relaxation we may consider a resonance experiment in which a microwave field is tuned to resonate with a simple two-level spin system. As discussed in Chapter 1 the induced transition probabilities for upward and downward transitions are equal (say W) so the upward and downward transition rates are given by $n_1 W$ and $n_2 W$, respectively, where n_1 and n_2 are the populations of the lower and upper levels. In thermal equilibrium n_1 and n_2 are related by

$$\frac{n_1}{n_2} = \exp\left\{\frac{\delta}{kT}\right\} \tag{2.8}$$

where $\delta = E_2 - E_1 = h\nu_0$.

Thus $n_1 > n_2$ and the rate of upward transitions induced by the microwave field is greater than the downward. There is a net absorption of energy by the spin system $(n_1 - n_2)Wh\nu_0$ which tends to equalise n_1 and n_2. This is opposed by the thermalising effect of spin–lattice relaxation which is trying to maintain n_1 and n_2 in the ratio of equation (2.8). Equilibrium is reached when the rate of absorption by the spins is exactly balanced by the rate of transfer to the lattice.

Formally, this may be expressed by writing rate equations for the level populations. For a two-level system:

$$\frac{dn_1}{dt} = -n_1(W+w_{12})+n_2(W+w_{21}) \tag{2.9}$$

$$\frac{dn_2}{dt} = -\frac{dn_1}{dt}$$

where w_{ij} is the probability for a transition from level i to level j induced by lattice vibrations. The upward probability w_{12} is not equal to the downward w_{21} as can be seen by putting $W = 0$ in equation (2.9). At thermal equilibrium $dn_1/dt = 0$ and we have:

$$\frac{w_{12}}{w_{21}} = \frac{n_2}{n_1} = \exp\left\{-\frac{\delta}{kT}\right\} \tag{2.10}$$

By the principle of detailed balance (Roberts, 1940; Kittel, 1958) a relation of this kind holds for all pairs of levels in a multilevel system also.

Equation (2.9) may be rewritten in terms of the population difference $\Delta n = n_1 - n_2$ and the total number of ions per unit volume $N = n_1 + n_2$ as follows:

$$\frac{d(\Delta n)}{dt} = N(w_{21}-w_{12})-\Delta n(2W+w_{12}+w_{21}) \tag{2.11}$$

Now suppose the level populations are disturbed by the application of a pulse of microwave power at the resonant frequency such that at the end of the pulse (time $t = 0$) the populations are equal, i.e. $\Delta n = 0$. Then in the absence of any perturbation ($W = 0$) Δn will recover to its thermal equilibrium value Δn_0 at a rate given by:

$$\Delta n = \Delta n_0 \{1-e^{-t/T_1}\} \tag{2.12}$$

where

$$\frac{1}{T_1} = (w_{12}+w_{21}) \text{ and } \Delta n_0 = N\left\{\frac{w_{21}-w_{12}}{w_{21}+w_{12}}\right\}$$

T_1 is known as the spin–lattice relaxation time which may typically have a value of order 100 msec at 4·2°K but usually becomes rapidly shorter as the temperature rises (see the paper by Pace *et al.* (1960) which is reproduced in Part II).

Returning again to equation (2.11) we see that the steady state solution in the presence of a microwave field ($dn/dt = 0$) implies:

$$\Delta n = \Delta n_0 \left[\frac{1}{1+2WT_1} \right] \tag{2.13}$$

Thus if $W \ll 1/T_1$, n remains very close to its thermal equilibrium value. On the other hand, if $W \gg 1/T_1$, n tends to zero and the transition is said to be saturated.

It is useful to have some idea of the amount of microwave power required to saturate a transition and this involves a knowledge of W. W is proportional to H_1^2 (where H_1 is the amplitude of the microwave magnetic field at the crystal) and inversely proportional to the width of the resonance line. In fact, for a two-level system:

$$W = \gamma^2 H_1^2 T_2 \quad \text{where} \quad \gamma = \frac{g\beta}{\hbar} \tag{2.14}$$

If we assume the paramagnetic sample to be placed in a microwave cavity with a magnification factor Q we can relate H_1 to the incident power P as follows:

$$P \approx \frac{\nu_0 V}{4Q_0} H_1^2 \tag{2.15}$$

where V is the effective volume of the cavity (i.e. $H_1^2 V = \int_{\text{cavity}} H_1^2 dV$). For $\nu_0 = 10^{10}$ Hz, $V \sim 1 \text{ cm}^3$ and $Q \approx 2500$ (2.15) gives $P \sim 10^6 H_1^2$ ergs/sec $\sim 0{\cdot}1\ H_1^2$ watts. The saturation condition may then be written:

$$P \gg \frac{1}{10\gamma^2 T_1 T_2} \tag{2.16}$$

and for $g = 2$, $T_1 \approx 10^{-1}$ sec and $T_2 \approx 10^{-8}$ sec we obtain $P \gg 3 \times 10^{-7}$ watts. Thus, a power level of about 10 μwatts should be sufficient to satisfy the condition $W \gg 1/T_1$ provided the sample is situated in a cavity resonant at the pumping frequency.

However, in a non-resonant situation, e.g. if the sample is in a waveguide, the effective Q may be some 10^3 times smaller and some tens of milliwatts required to achieve saturation.

Power levels of this order are readily available from several kinds of microwave source but it is apparent from equation (2.16) that shorter spin–lattice relaxation times or wider resonance lines may make more serious demands on pump power. This is one factor which limits practical masers to low temperature operation as T_1 tends to decrease rapidly with rising temperature. An example of this for Cr^{3+} in Al_2O_3 is provided by the measurements of Pace, Sampson and Thorp (1960) already referred to.

So far in this discussion of saturation we have assumed the resonance line to be homogeneously broadened so the line saturates uniformly. If, on the other hand, it is inhomogeneously broadened a very different situation may occur as the resonance condition is only satisfied for a small fraction of the ions in the crystal. In terms of Fig. 10 only a few of the individual lines which go to make up the profile will be saturated and the result may be that a hole is "burned" in the profile.

Such a situation is highly undesirable in a maser material as attempts to saturate the pump transition will only affect a small proportion of the active ions and maser gain will be correspondingly reduced. This is probably the reason why the Ni^{2+} ion has never been used successfully, though Bloembergen (1956) originally suggested Ni^{2+} in $ZnSiF_6–6H_2O$ as a likely material. This ion appears to be unusually sensitive to imperfections in the host crystal and its resonance lines are usually inhomogeneously broadened.

2.6. Inversion in a Multilevel System

In Chapter 1 we derived an expression for the inverted population difference in a three-level atomic system when saturating power is applied at the pump frequency ν_{13}. In practice most masers have employed the four levels of the $Cr^{3+}\,{}^4A_2$ ground state, but it is obviously possible to analyse this four-level system in a similar way. The general expression for the inverted population difference is cumbersome, but it is similar to that for the three-level case in depending on ratios between various spin–lattice coupling parameters w_{ij}. It is obviously desirable for understanding the behaviour of a particular material to know something of these coupling rates.

Their order of magnitude may be obtained from measurement of spin–lattice relaxation times, but the situation is rather more complex than for the two-level case considered above. Again we may write a set of rate equations having the form:

$$\frac{dn_i}{dt} = -n_i \sum_j (W_{ij}+w_{ij})+\sum_j n_j (W_{ji}+w_{ji}) \tag{2.17}$$

where

$$\sum_i n_i = N$$

If we assume the populations of a pair of levels to be disturbed by application of a saturating pulse of power at the appropriate resonance frequency it can be shown that the recovery of the population difference Δn is given (for a four-level system) by:

$$\Delta n = \Delta n_0 \{1-A_1 e^{-t/T_1}-A_2 e^{-t/T_2}-A_3 e^{-t/T_3}\} \tag{2.18}$$

where the time constants T_i depend on all the w_{ij} and the amplitudes A_1 depend on both the w_{ij} and the boundary conditions (i.e. the populations n_i at time $t = 0$). It is obvious from equation (2.18) that one can no longer define a unique spin–lattice relaxation time nor can one hope to obtain values for the individual w_{ij} from a single measurement of relaxation behaviour.

An attempt to overcome this difficulty was made by Orton *et al.* (1966) who calculated the w_{ij} from the theory of Van Vleck (1940) and used the calculated values to predict relaxation behaviour for Cr^{3+} in $ZnWO_4$. Satisfactory agreement with experimental results suggested these theoretical values to be approximately correct (for magnetically dilute material) but the accuracy is probably insufficient for making reliable inversion calculations. In practice, in assessing a maser material, the inversion must be measured experimentally. In any case there is another factor determining the behaviour of the material when the concentration of Cr^{3+} ions is in the region 10^{18}–10^{19}/cc (typical of a maser material). This is the phenomenon of harmonic cross relaxation first suggested by Bloembergen *et al.* (1959).

We have already outlined the way in which dipole–dipole coupling between similar paramagnetic ions may cause mutual "spin flips", such as illustrated in Fig. 9. For a two-level system this interaction results in line broadening but produces no net

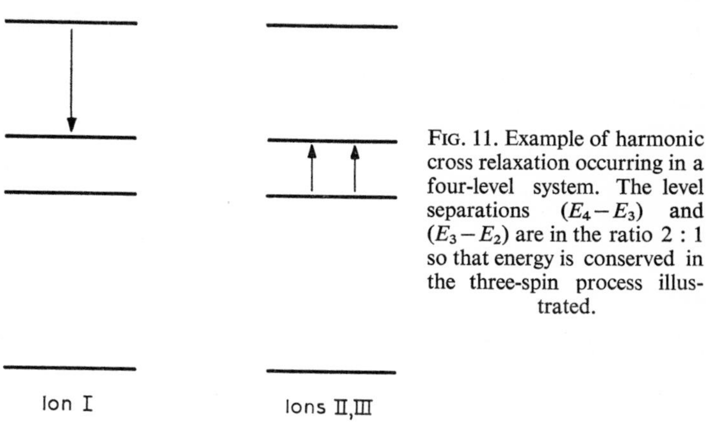

FIG. 11. Example of harmonic cross relaxation occurring in a four-level system. The level separations $(E_4 - E_3)$ and $(E_3 - E_2)$ are in the ratio 2 : 1 so that energy is conserved in the three-spin process illustrated.

change in the population distribution. This is no longer true for the case of a multilevel system and Fig. 11 shows an example of a "three-spin" process which may occur with conservation of energy when the harmonic relation $\nu_{34} = 2\nu_{23}$ exists between the level separations. It is clear that such a process does alter the net population distribution and its importance may be appreciated if we suppose a pump to be applied between levels 1 and 3 in an attempt to invert the populations of levels 2 and 3. This particular process will increase the inversion though obviously the reverse process will have the opposite effect.

To determine the net effect of harmonic cross relaxation one must include extra terms in the rate equations. For example, the process illustrated in Fig. 11 would require the addition of a term $n_4 n_2^2 W_{CR}$ to the equation for dn_3/dt while the reverse process would require the term $-n_3^3 W_{CR}$, where W_{CR} is the probability for the particular process. The transition rate is proportional to the probability of there being an ion in level 4 and two ions in level 2—hence the factor $n_4 n_2^2$. It is apparent that the rate equations involving cross relaxation are no longer linear in n_i which makes their solution much more difficult. In general it is only possible using numerical methods.

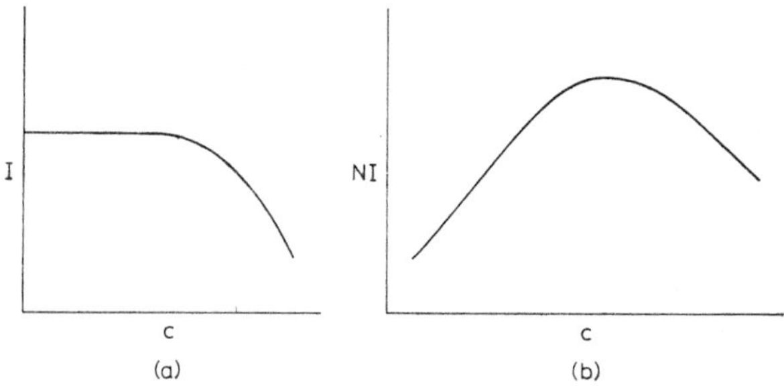

FIG. 12. Qualitative sketch of the variation of (*a*) the inversion I and (*b*) the NI product as functions of the active ion concentration C.

The transition probability W_{CR} depends on the probability of finding three paramagnetic ions within the range of their dipolar interactions which is proportional to C^2 (where C is their concentration). A cross-relaxation process involving N spins will be proportional to C^{N-1} so at low concentrations high-order processes will be negligible but as C increases they will become relatively more important. The many high-order processes present at concentrations of magnetic ions much above 10^{20}/cc (referred to as "general cross relaxation") tends to spread the pump saturation over all levels, thus equalising their populations and frustrating any attempt to achieve inversion. It is for this reason that the concentration cannot be increased indefinitely in a practical maser material. A measurement of inversion as a function of concentration yields a curve of the type shown in Fig. 12(*a*), being independent of C at low concentration and falling rapidly once general cross relaxation becomes significant. The parameter of importance for maser operation is the product NI which is shown in Fig. 12(*b*). It is obviously desirable to work at the concentration where this curve peaks.

The relative importance of cross relaxation and, consequently, the position of the NI maximum depends on temperature. It is apparent from a consideration of the rate equations that the significant parameter is the ratio of W_{CR} to a suitably averaged spin–lattice relaxation rate $\bar{w}_{ij}$ and, as has already been emphasised, the w_{ij} can depend strongly on temperature. W_{CR}, depending as it does on dipolar interaction between neighbouring magnetic ions, is more or less independent of temperature. Thus, the optimum concentration of ruby for a maser operating at 1·5°K is approximately 0·03 mole %, whereas Maiman (1960) found that at 77°K the NI maximum occurred at about 0·2 %.

Before closing this account of relaxation phenomena mention must be made of an experiment carried out in 1956 by Scovil, Feher and Seidel (1957) whose paper is reproduced in Part II. They demonstrated maser action in a crystal of lanthanum ethyl sulphate containing two paramagnetic ions Gd^{3+} and Ce^{3+}.

To understand the purpose of the double doping we must return to the expression for the inverted population in a three-level system (equation (1.22)). It is apparent that the inversion may be increased by using a pump frequency large compared with the signal frequency (implying $\nu_{idler} \gg \nu_{signal}$) or by making the idler relaxation rate large compared with that for the signal transition. Scovil *et al.* employed the latter method, using Ce^{3+} as a "cross-doping" ion to couple energy from the idler levels.

The principle of the experiment is that cross relaxation may occur between unlike ions provided they have appropriate level splittings and this process may proceed much more rapidly than spin–lattice relaxation. If the second ion is strongly coupled to the lattice it may therefore provide an effective "short circuit" between the maser ion and the lattice for the particular levels which are matched. This is illustrated in Fig. 13 for a three-level maser ion. It is clear that the cross doping ion will tend to maintain a Boltzmann distribution between levels 2 and 3 so when the pump transition is completely saturated the ratio n_2/n_1 is equal to

$$\exp\left\{\frac{E_3-E_2}{kT}\right\}$$

which is always positive no matter what frequency ratios are employed.

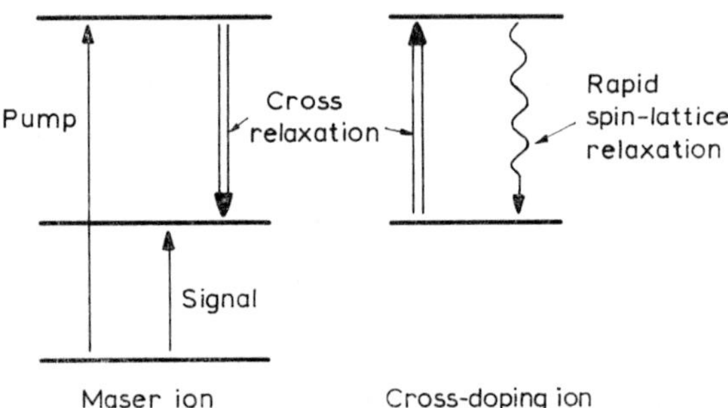

FIG. 13. Level diagram to illustrate the use of a "cross-doping" ion. When cross relaxation between the two ions is sufficiently rapid the "cross-doping" ion maintains a normal Boltzmann distribution between the idler level populations.

It is somewhat ironic that such a sophisticated experiment should be performed at a very early stage in the development of the solid state maser yet the technique has never been employed in practice. The reason lies in the difficulty of finding suitable cross-doping ions which can be "tuned" to match the appropriate maser levels.

2.7. Optical Pumping

For successful operation of a conventional three-level maser one requires a pump frequency considerably higher than the signal frequency and for a maser operating at signal frequencies much above 30 GHz it becomes increasingly difficult to find suitable pump sources. This prompts the thought that large inversions may be obtained by increasing the pump frequency into the optical range and using the high-power densities available from solid state lasers (Allen, 1968). Perhaps even more important is the possibility that optical pumping might allow useful C.W. maser operation at the temperature provided by liquid nitrogen (77°K) —which is more accessible than that of liquid helium.

There are two requirements implicit in this suggestion. The pump line width should be narrow enough to pump selectively from a single ground state level and the maser material should possess an excited state at the correct energy to absorb laser light (which must also be suitably sharp in frequency). Solid state laser line widths are usually small enough to satisfy the first requirement but the matching of energy levels provides a much more stringent condition. The tunability of a solid state laser is very small ($\sim 0{\cdot}1\%$) which makes it highly improbable that a chance coincidence of frequencies could be found. The only practical hope is to use the same material for both laser and maser when, provided the terminal level of the laser transition lies in the ground state, an automatic match is available.

It may be noted that the maser ion need no longer have a multilevel ground state as the third level is now an "optical" level but the list of possible materials is enormously reduced by the requirement of finding a laser with a ground state terminal level. Only two appear to be currently available, ruby and Tm^{2+} in CaF_2.

The possibility of using ruby is interesting in view of its suitability as a material for conventional maser operation and inversion in the Cr^{3+} ground state has been reported by Devor *et al.* (1962). They used a pulsed ruby laser and observed amplification and oscillation at a frequency of 22·4 GHz. Both laser and maser were operated at 4·2°K. Further details may be found in Part II. Similar results have also been reported by Zverev *et al.* (1964) who obtained oscillation at $\sim$30 GHz from an optically pumped ruby at 77°K but again only in pulsed operation.

There has been no report of C.W. operation of an optically pumped ruby maser and Walling (unpublished) has shown in a detailed analysis that pump power requirements for achieving useful C.W. operation appear to be beyond the capabilities of present ruby lasers even for operation at 4·2°K.

McLaughlan (1963) considered the possibility of pumping a Tm^{2+} CaF_2 maser with a laser using the same material and concluded that inversion of the Tm^{2+} ground state levels could be achieved provided the spin–lattice relaxation time of the ground state T_1 is longer than the fluorescent lifetime of the excited state τ. The value of τ has been measured as 4 msec while at helium temperatures Sabisky and Anderson (1966) find $T_1 \sim 1$ sec* so this system may be suitable. However, in their paper Sabisky and Anderson report a much simpler method of pumping the Tm^{2+} ground state using circularly polarised light to obtain preferential pumping of the lower level. This avoids the need for a laser and they achieved steady state inversion at a temperature of 1·5°K using a mercury lamp as pump source. High-temperature operation appears unlikely, however, as T_1 becomes rapidly shorter above about 8°K.

2.8. The Spin Hamiltonian

The papers by Geusic (1956) and Schulz-du Bois (1959) reproduced in Part II illustrate the use of the spin Hamiltonian in describing the energy levels used in a maser material. In Geusic's paper we see how the spin Hamiltonian parameters are determined from electron spin resonance measurements. Schulz-du Bois then describes the use of the spin Hamiltonian to calculate energy level splittings and transition probabilities with a view to choosing a suitable maser operating point.

To aid the understanding of these papers we now consider the spin Hamiltonian in rather more detail. As already described in

* N.B. There is a misprint in this paper: 700 μsec should read 700 msec. Since then these workers have measured values of T_1 as long as 20 sec (E. S. Sabisky, private communication).

the qualitative account given in Section 2.4 we consider a small group of ground state levels closely spaced amongst themselves, but separated by several hundred wave numbers (or more) from the next higher group. We can then define an effective spin S such that the multiplicity of the ground state, $n = 2S+1$. The magnetic quantum number M may take values $S, (S-1), \ldots (-S)$ and the set of numbers M defines a set of basic states (or wave functions) which may be written in the Dirac notation as $| M \rangle$. The energies of the $(2S+1)$ levels can then be obtained by solving the usual eigenvalue equations.

$$\mathscr{H} | \psi_i \rangle = E_i | \psi_i \rangle \quad i = 1, 2, \ldots (2S+1) \tag{2.19}$$

where the eigenfunctions $| \psi_i \rangle$ are, in general, linear combinations of the $| M \rangle$, i.e.

$$| \psi_i \rangle = a_i | M = -S \rangle + b_i | M = -S+1 \rangle + \ldots r_i | M = S \rangle \tag{2.20}$$

The operator $\mathscr{H}$ is the spin Hamiltonian for the particular system under consideration. It usually contains a number of terms representing the effect of the crystal field (where $S > \frac{1}{2}$) and the applied magnetic field H. Each term contains spin operators such as S_x, S_z^2, etc., the rules for operating on the states $| M \rangle$ being as follows (see, for example, Slater, 1960):

$$S_z \; | M \rangle = M | M \rangle \tag{2.21}$$

$$S_+ | M \rangle = \{S(S+1) - M(M+1)\}^{\frac{1}{2}} | M+1 \rangle \tag{2.22}$$

$$S_- | M \rangle = \{S(S+1) - M(M-1)\}^{\frac{1}{2}} | M-1 \rangle \tag{2.23}$$

where $S_\pm = S_x \pm iSy$.

The use of the spin Hamiltonian can probably be made clear most easily by treating a simple example. Let us take a single Kramers doublet for which $S = \frac{1}{2}$ and consider the effect of an applied magnetic field. The basic states for this case are $| M = \frac{1}{2} \rangle$ and $| M = -\frac{1}{2} \rangle$ and the Hamiltonian takes the form:

$$\mathscr{H} = \beta \boldsymbol{H} \, . \, g \, . \, \boldsymbol{S} \equiv \beta\{g_x H_x S_x + g_y H_y S_y + g_z H_z S_z\} \tag{2.24}$$

For simplicity we suppose H to be applied parallel to the z-axis

(z being determined by the symmetry of the crystal field) when $\mathscr{H} = g_z\beta HS_z$ and the eigenvalue equation reduces to the simple form:

$$\begin{aligned}\mathscr{H}\,|\pm\tfrac{1}{2}\rangle &\equiv g_z\beta HS_z\,|\pm\tfrac{1}{2}\rangle \\ &= \pm\tfrac{1}{2}g_z\beta H\,|\pm\tfrac{1}{2}\rangle\end{aligned} \tag{2.25}$$

where we have used equation (2.21). Comparing this with equation (2.19) shows that the energies of the two states $|\,M = \pm\frac{1}{2}\rangle$ are:

$$E_{\pm} = \pm\tfrac{1}{2}g_z\beta H \tag{2.26}$$

where

$$h\nu = E_{+}-E_{-} = g_z\beta H \tag{2.27}$$

which is the resonance condition described in equation (2.2).

The solution of the eigenvalue equations is less trivial if we suppose H to be applied parallel to the x-axis. The Hamiltonian becomes $\mathscr{H} = g_x\beta HS_x \equiv \frac{1}{2}g_x\beta H(S_{+}+S_{-})$ and operating with this on the states $|\pm\frac{1}{2}\rangle$ yields:

$$\mathscr{H}\,|\pm\tfrac{1}{2}\rangle = \tfrac{1}{2}g_x\beta H\,|\mp\tfrac{1}{2}\rangle \tag{2.28}$$

where we have used equations (2.22) and (2.23). Thus we obtain in the usual way (Pauling and Wilson, 1935) the secular equation:

$$\begin{vmatrix} -E & \tfrac{1}{2}g_x\beta H \\ \tfrac{1}{2}g_x\beta H & -E \end{vmatrix} = 0 \tag{2.29}$$

having solutions $E_{\pm} = \pm\frac{1}{2}g_x\beta H$ as might have been expected.

It is easy to show that the secular equation for $H \parallel x$ will be diagonal if instead of choosing states $|\pm\frac{1}{2}\rangle$ we take the linear combinations

$$|\,\psi_{\pm}\rangle = \frac{1}{\sqrt{2}}\{|\,\tfrac{1}{2}\rangle \pm |-\tfrac{1}{2}\}$$

i.e.

$$\mathscr{H}\,|\,\psi_{\pm}\rangle = E_{\pm}\,|\,\psi_{\pm}\rangle = \pm\tfrac{1}{2}g_x\beta H\,|\,\psi_{\pm}\rangle \tag{2.30}$$

This is a particularly simple example illustrating the general rule that the secular equation can always be diagonalised by the choice of a suitable linear combination of basic states, though in many cases there is no easy algebraic solution and numerical methods must be adopted.

We now consider the extension of these ideas to describe the four-fold degenerate Cr^{3+} ground state in a crystal field of axial symmetry such as is the case for ruby. An extra term must be included in the spin Hamiltonian to represent the effect of the crystal field and for an axially symmetric crystal the Hamiltonian is written:

$$\mathscr{H} = \beta \boldsymbol{H} . g . \boldsymbol{S} + D[S_z^2 - \tfrac{1}{3}S(S+1)] \tag{2.31}$$

where the effective spin of the Cr^{3+} ion is $S = 3/2$ giving a set of basic states $|\tfrac{3}{2}\rangle$, $|\tfrac{1}{2}\rangle$, $|-\tfrac{1}{2}\rangle$, and $|-\tfrac{3}{2}\rangle$. To see the effect of this new term let us first of all suppose $H = 0$ when the Hamiltonian reduces to $\mathscr{H} = D[S_z^2 - \tfrac{5}{4}]$. Operating on each of the states $|M\rangle$ twice with S_z gives $S_z^2 |M\rangle = M^2 |M\rangle$ and we readily obtain:

$$E(\pm\tfrac{3}{2}) = D$$

$$E(\pm\tfrac{1}{2}) = -D \tag{2.32}$$

showing that the four-fold degeneracy is partially removed, giving two Kramers doublets ($M = \pm\tfrac{1}{2}$) and ($M = \pm\tfrac{3}{2}$) separated by an amount $2D$.

If we now take the magnetic field to be applied parallel to the z-axis (i.e. the axis of crystal symmetry) the Hamiltonian becomes:

$$\mathscr{H} = g_z \beta H S_z + D[S_z^2 - \tfrac{5}{4}] \tag{2.33}$$

and the eigenvalues are:

$$E(\pm\tfrac{3}{2}) = D \pm \tfrac{3}{2} g_z \beta H$$

$$E(\pm\tfrac{1}{2}) = -D \pm \tfrac{1}{2} g_z \beta H \tag{2.34}$$

This is illustrated in Fig. 8(a): energy varies linearly with H and each level is described by one of the basic functions $|M\rangle$.

Now consider the more general case when H is applied at an angle θ to the z-axis as illustrated in Fig. 14. The spin Hamiltonian becomes

$$\mathscr{H} = g_{\parallel}\beta H_z S_z + g_{\perp}\beta(H_x S_x + H_y S_y) + D[S_z^2 - \tfrac{5}{4}] \qquad (2.35)$$

where $g_{\parallel} \equiv g_z$ and $g_x = g_y \equiv g_{\perp}$ (for axial symmetry $g_x = g_y$).

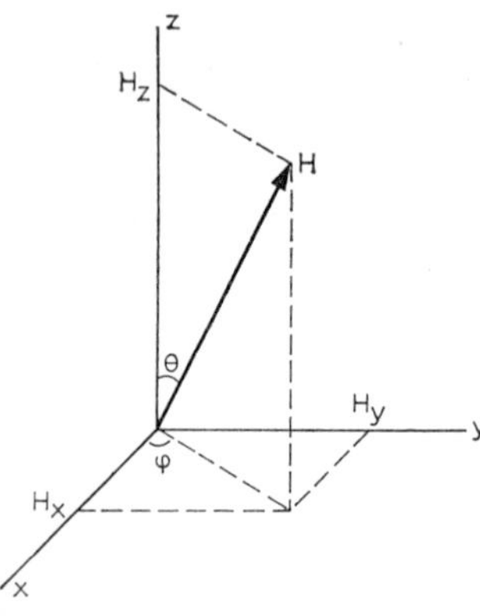

FIG. 14. System of axes to illustrate the orientation of the magnetic field with respect to the ruby crystal when calculating energies and transition probabilities. The Z direction coincides with the crystal c-axis.

Equation (2.35) may be written in terms of spherical polar coordinates as follows:

$$\begin{aligned}\mathscr{H} &= g_{\parallel}\beta H \cos\theta S_z + \tfrac{1}{2}H \sin\theta\,[\cos\varphi(S_+ + S_-) \\ &\quad - i\sin\varphi(S_+ - S_-)] + D[S_z^2 - \tfrac{5}{4}] \\ &= g_{\parallel}\beta H \cos\theta S_z + \tfrac{1}{2}H \sin\theta[e^{-i\varphi}\,S_+ + e^{i\varphi}\,S_-] + D[S_z^2 - \tfrac{5}{4}]\end{aligned} \qquad (2.36)$$

which is the form used by Geusic in his equation (2).

Making use of equations (2.22) and (2.23) to calculate matrix elements of the operators S_+ and S_- yields the secular determinant shown in Table 2, which is non-diagonal and the roots may only be determined by solving a quartic equation. There are, however, three special cases for which the determinant factorises: (1) $\theta = 0$ (the case considered above); (2) $\theta = \cos^{-1} 1/\sqrt{3} \approx 54{\cdot}7°$, where the energy levels are symmetrical about the $E = 0$

TABLE 2

	$\left\vert \frac{3}{2} \right\rangle$	$\left\vert \frac{1}{2} \right\rangle$	$\left\vert -\frac{1}{2} \right\rangle$	$\left\vert -\frac{3}{2} \right\rangle$
$\left\vert \frac{3}{2} \right\rangle$	$D+\frac{3}{2}g_{\parallel}\beta H \cos\theta - E$	$\frac{\sqrt{3}}{2}g_{\perp}\beta H \sin\theta e^{i\varphi}$		
$\left\vert \frac{1}{2} \right\rangle$	$\frac{\sqrt{3}}{2}g_{\perp}\beta H \sin\theta e^{-i\varphi}$	$-D+\frac{1}{2}g_{\parallel}\beta H \cos\theta - E$	$g_{\perp}\beta H \sin\theta e^{i\varphi}$	
$\left\vert -\frac{1}{2} \right\rangle$		$g_{\perp}\beta H \sin\theta e^{-i\varphi}$	$-D-\frac{1}{2}g_{\parallel}\beta H \cos\theta - E$	$\frac{\sqrt{3}}{2}g_{\perp}\beta H \sin\theta e^{i\varphi}$
$\left\vert -\frac{3}{2} \right\rangle$			$\frac{\sqrt{3}}{2}g_{\perp}\beta H \sin\theta e^{-i\varphi}$	$D-\frac{3}{2}g_{\parallel}\beta H \cos\theta - E$

axis; (3) $\theta = \pi/2$. For the latter case it is easy to show by expanding the determinant that the levels are given by:

$$E = \begin{cases} \frac{1}{2}G_{\perp} \pm \{G_{\perp}^2 - G_{\perp}D + D^2\}^{\frac{1}{2}} \\ -\frac{1}{2}G_{\perp} \pm \{G_{\perp}^2 + G_{\perp}D + D^2\}^{\frac{1}{2}} \end{cases} \tag{2.37}$$

where we have written $G_{\perp} \equiv g_{\perp}\beta H$. These equations are plotted in Fig. 8(*b*). The energy is no longer a linear function of H and the eigenstates are no longer "pure" but consist of linear combinations of the basic states $\vert M \rangle$ as previously discussed (Section 2.4).

For other angles the energies may only be calculated using numerical methods and Schulz-du Bois plots the results of such calculations for 10° intervals from $\theta = 0$ to 90°. (The results are independent of φ due to the axial symmetry of the problem.)

For the case of the Cr^{3+} ion in a crystal such as potassium cobalti-cyanide ($K_3Co(CN)_6$), where the site symmetry is orthorhombic, it is necessary to add yet another term to the spin Hamiltonian. This is usually written as

$$\mathscr{H}_{\text{rhombic}} = E[S_x^2 - S_y^2] \tag{2.38}$$

and introduces off-diagonal matrix elements into the secular determinant irrespective of the direction of H. This further complicates the calculation of the energies though analytic solutions are still possible for H parallel to the three principle directions x, y, z.

We shall not consider this case in any detail, though it was important for much of the early work on cavity masers using chromium-doped $K_3Co(CN)_6$.

For a complete analysis of a maser material we require knowledge not only of the energies of the allowed states but also the probabilities of transitions between them under the influence of an oscillatory magnetic field H_1.

In general the microwave field may be elliptically polarised, i.e. of the form:

$$\boldsymbol{H}_1 = \boldsymbol{H}' \sin \omega t + \boldsymbol{H}'' \cos \omega t \tag{2.39}$$

where $\boldsymbol{H}'$ and $\boldsymbol{H}''$ are mutually perpendicular vectors. Initially, however, we shall assume linear polarisation:

$$\boldsymbol{H}_1 = \boldsymbol{H}' \sin \omega t \tag{2.40}$$

It can be shown by applying time-dependent perturbation theory (e.g. Pake, 1962, but note that he defines $\boldsymbol{H}_1$ slightly differently) that the probability per unit time that an ion in the state $| M \rangle$ will make a transition to state $| M' \rangle$ is given by:

$$W_{MM'} = \frac{\pi^2 \beta^2}{h^2} | \langle M' | \boldsymbol{H}_1 . g . \boldsymbol{S} | M \rangle |^2 g(\nu) \tag{2.41}$$

where $g(\nu)$ is a line shape factor which peaks at $\nu = \nu_0 = (E_{M'} - E_M)/h$.

Taking the line shape to be the same for all transitions the transition probability depends only on the matrix element $\langle M' | \boldsymbol{H}_1 . g . \boldsymbol{S} | M \rangle$ and if, for simplicity, we assume g to be isotropic we can write

$$W_{MM'} \propto | \boldsymbol{H}_1 |^2 | \langle M' | \boldsymbol{h} . \boldsymbol{S} | M \rangle |^2 \tag{2.42}$$

showing the expected dependence on $| H_1 |^2$, i.e. on the microwave power falling on the crystal. $\boldsymbol{h}$ is a unit vector in the direction of $\boldsymbol{H}_1$.

We can illustrate the use of equation (2.42) by applying it to the simple case of an isolated doublet in a magnetic field and calculate the probability of transitions between the states $| M = -\frac{1}{2} \rangle$ and $| M = \frac{1}{2} \rangle$. The operator S_z has no matrix elements between these states (see equation (2.21)) so transitions are only induced by a component of $\boldsymbol{H}_1$ perpendicular to z. As the x-direction for this simple case is determined by the applied steady field $\boldsymbol{H}$ we see that we require a component of $\boldsymbol{H}_1 \perp \boldsymbol{H}$ and the maximum probability occurs when $\boldsymbol{H}_1 \perp \boldsymbol{H}$. Note also that the operators S_x and S_y only couple states differing by unity in the quantum number M, giving rise to the selection rule $\Delta M = \pm 1$ referred to previously.

For the case of the Cr^{3+} ground state where $S = 3/2$ the situation is rather more complicated. As explained above the states are, in general, "mixed" and transitions may occur between any pair of levels. The special case of axial symmetry with $H \| z$ is, of course, an exception, the $\Delta M = \pm 1$ selection rule again holds and transitions are induced only by components of $\boldsymbol{H}_1$ which are perpendicular to $\boldsymbol{H}$.

When $\boldsymbol{H}$ is not parallel to z it is still possible to calculate transition probabilities provided the exact form of the states is known, i.e. once we know the parameters a_i, b_i, etc., in the appropriate eigen-functions:

$$| \psi_i \rangle = a_i | \tfrac{3}{2} \rangle + b_i | \tfrac{1}{2} \rangle + c_i | -\tfrac{1}{2} \rangle + d_i | -\tfrac{3}{2} \rangle \tag{2.43}$$

These may also be calculated numerically once the energies have been determined and Schulz-du Bois plots values for all four levels as a function of H at 10° intervals of θ. (Note that his nomenclature is rather different from ours. His $\alpha(\bar{n}, m)$ is the coefficient of the state $| M = m \rangle$ in the expansion of the function $| \psi_n \rangle$, i.e.

$$| \psi_n \rangle = \alpha(\bar{n}, \tfrac{3}{2}) \, | \tfrac{3}{2} \rangle + \alpha(\bar{n}, \tfrac{1}{2}) \, | \tfrac{1}{2} \rangle + \alpha(\bar{n}, -\tfrac{1}{2}) \, | -\tfrac{1}{2} \rangle$$
$$+ \, \alpha(\bar{n}, -\tfrac{3}{2}) \, | -\tfrac{3}{2} \rangle.)$$

Suppose the steady magnetic field is applied in the zx-plane so that the parameters a_1, etc., are real and we wish to determine the probability for the transition between states $| \psi_n \rangle$ and $| \psi_m \rangle$ induced by a linearly polarised microwave field. We must evaluate the element $\mu = \langle \psi_m | \boldsymbol{h} . \boldsymbol{S} | \psi_n \rangle$. Expanding $\boldsymbol{S}$ into components:

$$\boldsymbol{S} = S_x + S_y + S_z = \tfrac{1}{2}(S_+ + S_-) + i/2(S_+ - S_-) + S_z \quad (2.44)$$

this matrix element can be seen to have the form:

$$\mu = \boldsymbol{h} . (\alpha\hat{x} + i\beta\hat{y} + \gamma\hat{z}) \quad (2.45)$$

where $\hat{x}$, $\hat{y}$ and $\hat{z}$ are unit vectors parallel to the three principal directions and α, β, γ may be calculated from a_n, b_n, a_m, b_m . . ., etc. Equation (2.45) can be written:

$$\mu = \boldsymbol{h} . (\boldsymbol{A} + i\boldsymbol{B}) \quad (2.46)$$

where $\boldsymbol{A}$ and $\boldsymbol{B}$ are mutually perpendicular, $\boldsymbol{B}$ lying parallel to y and $\boldsymbol{A}$ in the zx-plane. The maximum transition probability occurs when $| \mu |^2 \equiv | \boldsymbol{h} . (\boldsymbol{A} + i\boldsymbol{B}) |^2$ is maximised. It is easy to see that the required condition is that $\boldsymbol{h}$ (and therefore $\boldsymbol{H}_1$) should be parallel to the larger of $| \boldsymbol{A} |$ or $| \boldsymbol{B} |$ giving $| \mu |^2 = | A |^2$ or $| B |^2$ as the case may be.

In practice it is usual for the microwave field to be elliptically polarised (equation (2.39)) and one must then maximise $| (\boldsymbol{H}' \pm i\boldsymbol{H}'') . (\boldsymbol{A} + i\boldsymbol{B}) |^2$. In general this is rather a difficult case to deal with but for circular polarisation ($H' = H''$) it is straightforward to show that the maximum value of $| \mu |^2$ is $H_1^2(A \pm B)^2$ and occurs when $\boldsymbol{H}'$ is parallel to $\boldsymbol{A}$ and $\boldsymbol{H}''$ is parallel to $\boldsymbol{B}$ (the negative sign being taken when A and B have opposite sign).

For optimum transition probability it can be shown (Section 3.4) that $| \mu |^2 \propto (A^2 + B^2)$.

CHAPTER 3

Maser Devices

3.1. Choice of Active Material

We have already discussed some aspects of Bloembergen's basic paper on the three-level solid state maser in which some of the factors affecting the choice of material are considered. The main criteria for a solid state maser material as already indicated in Chapter 2 are as follows:

1. The material must contain an ion whose ground state is split by the action of the internal electric crystal field and an external magnetic field into three or more components.
2. Stimulated transitions between the various components of the ground state must be allowed.
3. If possible all the paramagnetic ions in the system should take part in the interaction with the pump and signal fields.
4. Again in order to employ all the active ions the material should be one in which the paramagnetic resonance lines are not inhomogeneously broadened.
5. Three level maser operation depends on saturation of the pump transition. The spin lattice relaxation time should then be long, although we must recognise that if the pump transition is readily saturable then the probability is that so too will be the signal transition and this will have an adverse effect on the dynamic range of a maser device.
6. In order to make use of the maser material it is necessary to incorporate it in some microwave circuit—this is made easier if the dielectric constant of the material is not too

large or strongly temperature dependent. Therefore secondary criteria for a maser material are that it should not have too large a dielectric constant and that the dielectric constant should not be strongly temperature dependent.

7. Lastly, but in practice by no means least, a maser material must be readily obtainable in large single crystals of a high degree of perfection (the potentialities of at least one promising material—emerald—have not been fully exploited because of difficulty in realisation).

These requirements severely restrict the number of materials which can be used in maser devices, synthetic ruby (chromium-doped aluminium oxide) satisfies most of the criteria and has been widely used; however, it has rather a low zero field splitting which makes it unsuitable for applications at millimetre wavelengths and the majority of maser devices operating in the millimetre range have made use of rutile (TiO_2) containing either chromium (Cr^{3+}) or iron (Fe^{3+}) as the active ion. A major disadvantage of rutile is that it has a high, anisotropic and strongly temperature-dependent dielectric constant which makes it exceedingly difficult to incorporate in a controlled way into electrical structures—particularly those of the type used in travelling wave masers. Other materials which have been used in solid state masers are gadolinium (Gd^{3+}) in lanthanum ethyl sulphate, chromium (Cr^{3+}) in potassium cobalti-cyanide, chromium (Cr^{3+}) in zinc tungstate and emerald where the active ion is again Cr^{3+} in a beryllium aluminium silicate host.

The choice of a material for practical masers usually lies between ruby (Cr^{3+} in Al_2O_3) and rutile containing either Cr^{3+} or Fe^{3+}, although as we have pointed out rutile is less than an ideal material—the choice depends largely on the frequency of operation which is specified. For frequencies up to 10 GHz ruby with a zero field splitting of 11·46 GHz has proved very useful. Operation at higher frequencies calls for high magnetic fields and saturation of the pump transition becomes difficult as states become increasingly pure at high fields. In other materials, particularly rutile, the zero field splitting is substantially larger than in ruby (~45 GHz) and although high frequency operation may still call for high magnetic fields the states of the ground state levels are still mixed and the pump transitions remain strongly allowed.

Having chosen the active material it is then necessary to choose concentration and orientation in order to secure the maximum rate of emission of energy from the maser material for a given power at the signal frequency incident on the crystal. The rate of emission of energy from the maser crystal is given by

$$P_m = \Delta n V_c h\nu_s W_{MM'} \tag{3.1}$$

when Δn is the inverted population density difference between the signal transition levels, V_c the volume of the crystal, $h\nu_s$ the signal quantum and $W_{MM'}$ the transition probability at the signal frequency. Now as we have seen $W_{MM'}$ depends on the square of the r.f. magnetic field at the signal frequency (equation (2.41)) and thus P_m itself is not a useful figure of merit. It is more convenient to speak in terms of a quality factor Q_m defined as

$$Q_m = \frac{2\pi\nu_s W_s}{P_m} \tag{3.2}$$

Here W_s is the total stored energy in the cavity, in the case of a cavity maser or per unit length if we are dealing with a T.W. maser.

The transition probability $W_{MM'}$ is not concentration-dependent except insofar as it involves the line shape (see Section 2.8) and since the line shape is found to vary only slowly with concentration in the range of interest for practical purposes we may say that $W_{MM'}$ is independent of the concentration of active ions in the crystal. On the other hand, it varies strongly with orientation and to illustrate this point we have plotted in Fig. 15 the variation of the probability of the 1-2 transition in ruby at 4170 MHz with orientation (i.e. the angle which the d.c. magnetic field makes with the threefold axis of symmetry of the crystal). In plotting Fig. 15 the optimum value of transition probability has been taken in each case.

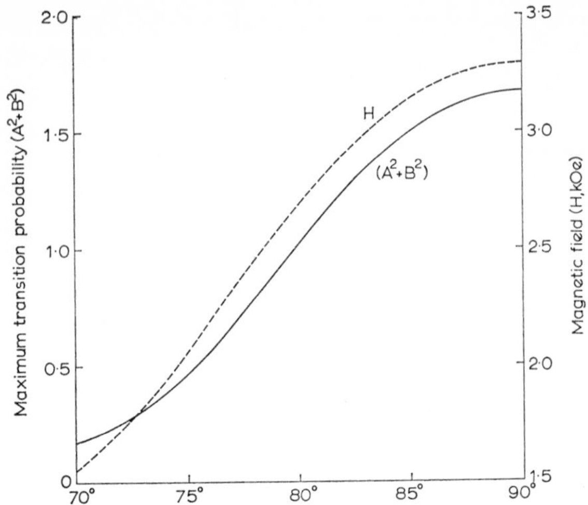

FIG. 15. Variation of the optimum transition probability (A^2+B^2) and applied field (H) for the 1–2 transition in ruby (Cr^{3+} in Al_2O_3) at 4170 MHz.

The inverted population difference Δn can be expressed (cf. Chapter 1) as

$$\Delta n = -\Delta n_0 I \tag{3.3}$$

Δn_0 being the equilibrium population difference and I the inversion produced by a particular pumping scheme. Δn_0 of course, depends linearly on concentration and I is, at very low concentrations, concentration-independent being determined by the various transition frequencies and spin–lattice interaction processes. We recall (Chapter 1) that for a three level system the inversion of levels 1 and 2 which results from application of a pump between levels 1 and 3 is given by

$$I = -\frac{\Delta n}{\Delta n_0} = \frac{w_{23}\,\nu_{23} - w_{12}\,\nu_{12}}{w_{23}\,\nu_{12} + w_{12}\,\nu_{12}} \tag{3.4}$$

At high concentration the onset of general cross relaxation, i.e. multiple spin–spin energy exchange processes tends to preclude the establishment of the thermal non-equilibrium within the spin system and inversion thus decreases with concentration.

As we have already seen rather little is known about the individual spin–lattice interaction rates w_{ij}, and it has frequently been assumed that they are of comparable magnitude. This assumption is very questionable but if we accept it then the inversion depends simply on the ratio of the pump and signal frequencies

$$I = \frac{\nu_{13}}{2\nu_{12}} - 1 \tag{3.5}$$

The maser designer thus tends to choose an operating point which gives him as large a pump frequency as is practicable. This procedure is probably valid even if the w_{ij}, are not equal. On the other hand, it is possible in situations when ν_{23} is very close to ν_{12} to increase the inversion by artificially increasing w_{23}. This procedure was in fact followed by Scovil *et al.* (1957) in their highly sophisticated first demonstration of three-level maser operation. We shall consider their work in detail shortly but first let us spend some time on maser devices.

3.2. Cavity Masers

Maser operation depends on the interaction between the spin system in the paramagnetic maser crystal and the r.f. magnetic fields. The strength of this interaction is determined by the square of the amplitude of these fields within the crystal and therefore it is desirable that, for a given signal power input to the maser, this amplitude should be made as large as possible.

The simplest way of ensuring that a large r.f. field acts upon the crystal is to place it in a cavity resonant at the signal frequency. Such a device is referred to as a cavity maser and as mentioned in Chapter 1 all the early masers were of this type. It can be constructed as a reflection device (Fig. 3) in which input and output are separated by means of a circulator or as a transmission device with distinct input and output terminals. In either case the device is regenerative.

3.2.1. *Gain and bandwidth of a cavity maser*

The performance of a reflection cavity maser may be rather easily calculated using standard cavity theory. We will define the following quantities:

Q_0 = intrinsic cavity $Q = \dfrac{2\pi \nu_s W_s}{P_0}$; P_0 = power absorbed in the cavity walls

Q_e = external $Q = \dfrac{2\pi \nu_s W_s}{P_e}$; P_e = power coupled out of the cavity into the load

Q_m = magnetic $Q = \dfrac{2\pi \nu_s W_s}{P_m}$; P_m = power *absorbed by* the maser material

Here ν_s is the cavity resonant frequency and W_s the mean stored energy. Defining

$$\frac{1}{Q_c} = \frac{1}{Q_0} + \frac{1}{Q_m} \tag{3.6}$$

the input impedance of the cavity at resonance is given by

$$Z_c = \frac{Q_c}{Q_e} \tag{3.7}$$

and the voltage reflection coefficient by

$$\rho = \frac{Z_c - 1}{Z_c + 1} \tag{3.8}$$

Thus the power gain of the device is

$$G = \rho^2 = \left(\frac{Q_c - Q_e}{Q_c + Q_e}\right)^2 \tag{3.9}$$

The bandwidth to 3 dB points is given by the operating frequency divided by the loaded Q (Q_L; $1/Q_L = 1/Q_c + 1/Q_e$) and is

$$B = \frac{\nu_s}{Q_L} = \nu_s\left(\frac{1}{Q_c} + \frac{1}{Q_e}\right) \tag{3.10}$$

thus

$$G^{\frac{1}{2}}B = \nu_s\left(\frac{1}{Q_e}-\frac{1}{Q_c}\right) \tag{3.11}$$

It is not strictly true to state that the root gain bandwidth product $G^{\frac{1}{2}}B$, of a cavity maser is a constant; however, in practice it is usually the case that $Q_0 \gg Q_m$; furthermore, Q_m is negative when maser action is occurring and if large gain is to be obtained the coupling is adjusted so that $Q_e \doteqdot Q_m$. In these circumstances

$$G^{\frac{1}{2}}B \doteqdot \frac{2\nu_s}{|Q_m|} \tag{3.12}$$

This analysis is over-simplified particularly insofar as it neglects the frequency dependence of the paramagnetic resonance process. A more sophisticated analysis (Butcher, 1958) gives the same result for the centre frequency gain (it is assumed that the resonant frequency of the cavity is exactly equal to the transition frequency) but the expression for the gain bandwidth product is modified and becomes, under the condition $Q_0 \gg Q_m$, $Q_e \doteqdot (Q_m)$

$$(G-2)^{\frac{1}{2}}B = \frac{2\nu_s}{Q_m+Q_s} \tag{3.13}$$

where Q_s is the effective Q of the paramagnetic resonance line and is

$$Q_s = \frac{\nu_s}{\Delta\nu}; \tag{3.14}$$

$\Delta\nu$ being the width of the paramagnetic resonance line.

Armed with the expression for the gain of a cavity maser it is useful to follow the sequence of events in the operation of the device. We will distinguish two cases and illustrate them with sketches of a klystron mode reflected from the maser cavity.

(i) *Cavity initially over coupled, i.e.* $Q_0 > Q_e$ (Fig. 16)

(*a*) No magnetic field—off magnetic resonance—only the cavity resonance is visible.

(*b*) Resonant field applied; absorption reduces the net cavity Q and the reflection coefficient decreases—it may go through zero.

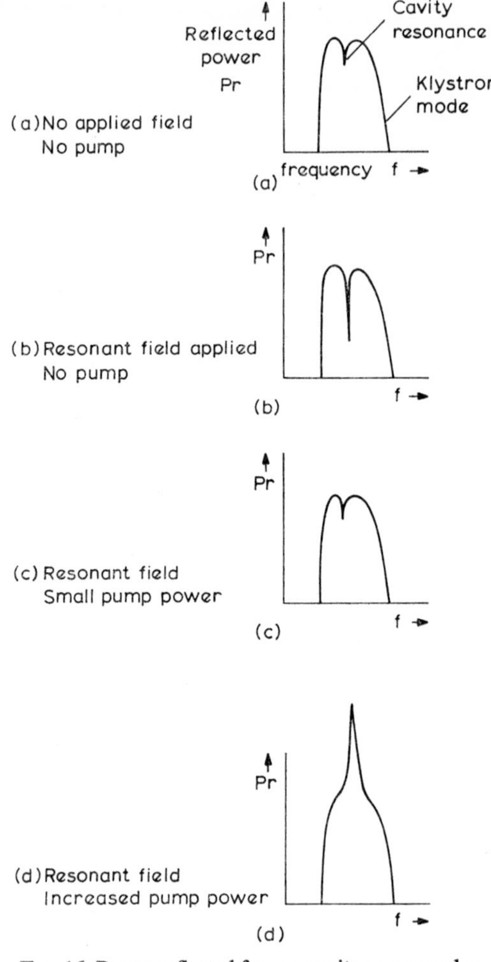

FIG. 16. Power reflected from a cavity maser under various conditions of applied field and pump power. Cavity initially over coupled.

(*c*) Pump frequency applied, crystal becomes transparent $Q_m \sim \infty$.

(*d*) As the pump power is increased the reflection coefficient increases passing through unity when $|Q_m| = Q_0$ and we have the characteristic "gain spike".

(ii) *Cavity initially under coupled, i.e.* $Q_0 < Q_e$ (Fig. 17)

(*a*) No magnetic field.

(*b*) Resonant field applied, reflection coefficient increases but does not reach unity.

(*c*) Pump power applied, reflection coefficient initially decreases rapidly passes through zero and then increases again giving the gain spike. Oscillation occurs if $Q_m < Q_e$ and this is more likely to happen in the under coupled case than in the over coupled case.

The performance of the cavity maser depends rather critically on the parameter Q_m. The determination of Q_m is discussed at the end of this chapter (Section 3.4) and it is shown that using ruby

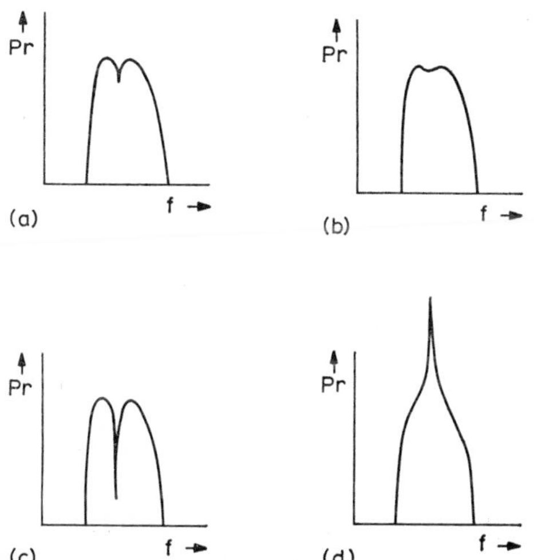

FIG. 17. As Fig. 16 with the cavity initially under coupled.

as the active material and assuming a filling factor of 0·3 (i.e. 30% of the total r.f. magnetic energy stored in cavity is stored in the maser crystal) a value of Q_m of 100 can be realised at 4·0°K. Actually it is possible in fully filled cavities to obtain filling factors close to unity and in such a case values of Q_m between 30 and 40 are achievable. The natural width of the paramagnetic resonance line in ruby (0·035% Cr : Al) is close to 63 MHz and thus at a frequency of 4200 MHz (the frequency taken in our calculation of Q_m) we have

$$Q_s = \frac{\nu_s}{\Delta\nu} = \frac{4200}{63} = 68$$

In this case then the gain bandwidth product of a reflection cavity maser will be given by

$$(G-2)^{\frac{1}{2}}B = \frac{8400}{30+68}\ \text{MHz} = 84\ \text{MHz}$$

Such a maser giving a gain of 20 dB will have a bandwidth of 8·4 MHz.

Performance figures such as these are about the best which could be achieved with a single cavity maser and the figures obtained in the early work, e.g. McWhorter and Meyer (1958), Makhov *et al.* (1958), are understandably much less impressive.

3.2.2. *The noise temperature of the cavity maser*

The most important practical attribute of the solid state maser is that it contributes very little noise to the signal which is being amplified, it is the nearest approach to the noiseless amplifier yet realised and indeed it is unlikely to be improved upon. It is worth while then spending a few moments in considering the noise characteristics of the device. We can say at the outset that the great advantage the maser has over the electron tube is that shot noise is entirely absent because in the maser we are concerned with the interaction of a r.f. magnetic field and localised paramagnetic ions. The only sources of noise in the device are essentially thermal and to understand these we must remind ourselves of a basic result of radiation theory which is that a body at a temperature T which absorbs a fraction α of radiant energy at a frequency ν incident upon it will re-radiate noise power into the band $\Delta\nu$ according to

$$P_m = \frac{\alpha h\nu\Delta\nu}{\exp(h\nu/kT)-1} = \alpha\varphi(T)\Delta\nu \qquad (3.15)$$

This expression assumes that the lossy element or body radiates into a system which can only support one mode at the frequency in question. It is appropriate to use it for non-degenerate systems. This expression may also be applied to systems characterised by negative temperatures (Chapter 1).

As we have seen it is possible in a multilevel system to produce a stable population distribution in which $n_2 > n_1$ and such a distribution is characterised by a negative spin temperature. This negative temperature is significant in so far as the contribution made by the maser material to the noise output of the maser is determined by it, the occurrence of negative temperatures may be rather startling but it is simply a convenient way of characterising a population distribution and is not related to the physical temperature of the material. The maser material will then radiate noise power at frequency ν into a band $\Delta\nu$ of

$$P_s = \alpha_s\varphi(T_s)\Delta\nu \qquad (3.16)$$

Clearly if T_s, the spin temperature, is negative the material emits energy at frequency ν, and if α_s is also negative, it follows that at all spin temperatures the maser material will emit noise radiation. Physically this is very reasonable, the origin of the noise radiation being spontaneous emission from the upper level of the two between which the signal transition occurs and, indeed, our expression for the noise power radiated from the maser material can readily be deduced on the basis of Einstein's relationship between the spontaneous and stimulated emission coefficients.

To determine the noise power output of the maser then we simply have to assign absorption coefficients and temperatures to the various parts and add their contributions. Considering the reflection cavity maser sources of output noise are as follows:

(i) loss in input and output lines;

(ii) losses in the cavity walls, etc.;

(iii) spontaneous emission from the maser material.

The effect of loss in the input and output lines is readily calculated. If the absorption coefficient is α we have, considering an element of length dz at a temperature T,

$$dP_N = -P_N \alpha dz + \alpha \varphi(T) \Delta \nu \, dz \tag{3.17}$$

If α and T are independent of z this equation can readily be integrated, in general both quantities will depend on z but it will in most cases be a sufficient approximation to take α and T as being constant and to use mean values. We then find that the noise power at the end of such a line P_{N_2} is related to the input noise power by

$$P_{N_2} = P_{N_1}(1-\lambda) + \lambda \varphi(T) \Delta \nu$$

where we have written

$$\lambda = 1 - e^{-\alpha l};$$

l being the length of the line. The effective absorption coefficients of the cavity and maser material are obtained as follows. We have

$$\frac{1}{Q_c} = \frac{1}{Q_0} + \frac{1}{Q_m} = \frac{P_0 + P_m}{2\pi \nu_s W_s} = \frac{P_i(1-G)}{2\pi \nu_s W_s} \tag{3.18}$$

(this is because the net power dissipated in the cavity and maser material is equal to the difference between incident power P_i and reflected power P_r and $P_r/P_i = G$).

$$\therefore \frac{P_0}{P_i} = \frac{Q_c}{Q_0}(1-G) = \text{absorption coefficient of the cavity walls (note that this is } \textit{always} \text{ positive)}, \tag{3.19}$$

and

$$\frac{P_m}{P_i} = \frac{Q_c}{Q_m}(1-G) = \text{absorption coefficient of the maser material (this may be positive or negative)}. \tag{3.20}$$

Assuming then that the gain of the maser is constant over the frequency interval $\Delta\nu$ the output noise power of the maser is

$$P_N = \Delta\nu\left\{\left[\underset{\text{(input line loss)}}{(\varphi(T_i)\,(1-\lambda)+\lambda(T_\lambda))G}+\underset{\text{(cavity losses)}}{\frac{Q_c}{Q_0}\,(1-G)\varphi(T_c)}\right.\right.$$

$$\left.\left.+\underset{\text{(maser material)}}{\frac{Q_c}{Q_m}\,(1-G)\varphi(T_s)}\right]\underset{\text{(output line)}}{(1-\lambda)+\lambda\varphi(T_\lambda)}\right\} \qquad (3.21)$$

here we have characterised the input noise power by the temperature T_i, and written T_λ for the mean temperatures of the input line, T_c the temperature of the cavity and T_s the spin temperature of the maser material. In practice we may write

$$\varphi(T) = kT \text{ — true for } \frac{h\nu}{kT} \ll 1$$

and then, bearing in mind that for an operating maser giving positive gain Q_c, Q_m and T_s are negative quantities, we find that the noise factor of the device

$$F = \frac{\text{output noise power/input noise power}}{\text{net gain}} \qquad (3.22)$$

is given by

$$F = 1+\frac{\lambda T_\lambda}{(1-\lambda)T_i}+\frac{(G-1)}{(1-\lambda)T_iG}\left[\frac{|Q_0|}{Q_0}\,T_c+\frac{|Q_c|}{|Q_m|}\,|T_s|\right]$$
$$+\frac{\lambda T_\lambda}{G(1-\lambda)^2T_i} \qquad (3.23)$$

For a high-gain device this simplifies to

$$F = 1+\frac{1}{(1-\lambda)T_i}\left[\lambda T_\lambda+\frac{|Q_c|}{Q_0}\,T_c+\frac{|Q_c|}{|Q_m|}\,|T_s|\right] = 1+\frac{T_m}{T_i} \qquad (3.24)$$

whence T_m, the equivalent noise temperature of the maser, is given by

$$T_m = \frac{1}{1-\lambda}\left[\lambda T_\lambda+\frac{|Q_c|}{Q_0}\,T_c+\frac{|Q_c|}{|Q_m|}\,|T_s|\right] \qquad (3.25)$$

To obtain a low noise temperature it is thus essential to minimise loss in the input waveguides and to employ a cavity with a very high intrinsic Q. A good approximation to T_m is then

$$T_m = \frac{1}{1-\lambda}\left[\lambda T_\lambda + |T_s|\right] \tag{3.26}$$

To obtain a low device noise temperature it is clearly necessary to reduce the losses in the input line to as low a value as possible. The spin temperature can readily be determined from a knowledge of the inversion since

$$(\Delta n)_0 = n_{10}(1-e^{-h\nu_s/kT_c}) = n_{10}\frac{h\nu_s}{kT_c}$$

$$(\Delta n) = n_1(1-e^{-h\nu_s/kT_s}) = n_1\frac{h\nu_s}{kT_s}$$

$$\therefore \frac{\Delta n}{(\Delta n)_0} = -I = \frac{n_{10}}{n_1}\frac{T_s}{T_c} \doteqdot \frac{T_s}{T_c} \quad \therefore T_s = -\frac{T_c}{I}$$

In practical masers it is possible to obtain inversions of 3 (cf. Section 3.4) and in such a case when the ambient temperature is 4·2°K the spin temperature T_s will be $-1{\cdot}4$°K. With careful attention to the design of the input lead its losses can be reduced to a few tenths of a decibel. If the lead loss is 0·25 dB, $\lambda = 0{\cdot}055$ and assuming the mean temperature of the lead T_λ to be 100°K we have $\lambda T_\lambda = 5{\cdot}5$°K.

Thus for this case

$$T_m = \frac{1}{0{\cdot}945}(5{\cdot}5+1{\cdot}4)$$

$$= 7{\cdot}3°\text{K}$$

It will be clear to the reader that the maser's extremely low noise temperature can be realised only if the most painstaking attention is given to the design of the input lead in order to reduce its losses and lower its temperature without adding appreciably to the consumption of refrigerant.

3.2.3. *Dynamic range of the cavity maser*

An important characteristic of any amplifier is the range of input powers over which it will give useful amplification. The lower limit to the dynamic range is determined by the bandwidth and equivalent noise temperature of the device; it is $kT_m\Delta\nu$. The upper limit is set by the gain saturation which occurs in the maser as a consequence of the decrease in the population difference between the signal levels resulting from the stimulated transitions (equation (1.21)) and it can readily be shown that the variation of gain (G) with input power is given by

$$\frac{(G^{\frac{1}{2}}+1)\,(G_0^{\frac{1}{2}}-1)}{(G^{\frac{1}{2}}-1)\,(G_0^{\frac{1}{2}}+1)}-1+\frac{P_i}{P_s}(G-1) = 0 \tag{3.27}$$

In this expression G is the numerical gain at an input signal level P_i, G_0 is the small signal numerical gain and P_s for a three-level system is given by

$$P_s = \Delta n' V_c h\nu_s\,(w_{12}+w_{23}) \tag{3.28}$$

$\Delta n'$ is the small signal population density difference between the signal levels, V_c the volume of the crystal and w_{ij} are the thermal

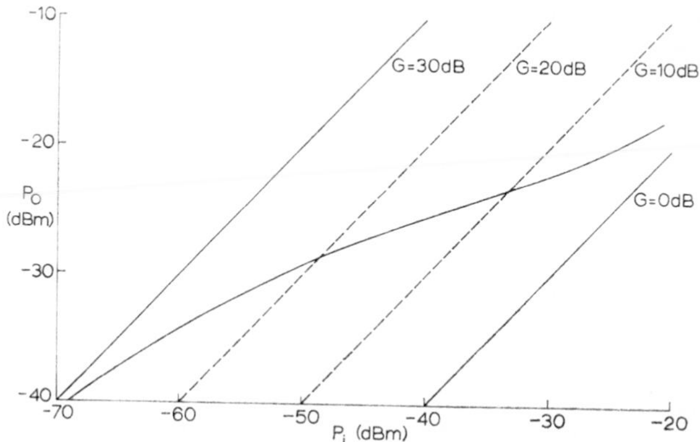

FIG. 18. Gain saturation in a cavity maser giving a small signal gain of 30 dB (calculated from equation (3.27) with $P_s = -20$ dBm).

relaxation rates already discussed in the preceding chapters. Physically P_s is the maximum rate at which energy can be emitted from the maser crystal at the frequency ν_s. Typically for a ruby maser operating at liquid helium temperatures P_s is about -20 dBm (10^{-5} watt).* In Fig. 18 we plot the dynamic characteristic of a reflection cavity maser giving a small signal gain of 30 dB with $P_s = -20$ dBm.

3.2.4. *Cavity maser devices*

The first successful demonstration of three-level solid state maser operation was by Scovil *et al.* (1957) and this demonstration was closely followed by the publication of details of a cavity maser operating at S-band (*ca.* 3000 MHz) by McWhorter and Meyer (1958) and these papers are reprinted in Part II of this volume. Subsequently many cavity masers have been described, among the more notable being the first ruby maser described by Makhov *et al.* (1958), the L-band (*ca.* 1300 MHz) device due to Artman *et al.* (1958) and the very high frequency ($\nu_s = 96$ GHz) devices using iron doped rutile, described by Foner *et al.* (1961) and Hughes (1962).

Generally, however, cavity masers have not been widely employed in systems and some of the disadvantages against their use in practical systems are as follows. In the first place the device is essentially regenerative and therefore tends to be unstable at high gains. Secondly, because of the constant root gain bandwidth condition it is not generally possible to secure operation over a wide band. To some extent the bandwidth restriction can be alleviated by employing reactance compensation as discussed by Kyhl *et al.* (1962) or multiple cavities, Nagy and Friedman (1962). Thirdly, the use of external non-reciprocal elements to separate input and output is necessary and, although these can sometimes be cooled, they always have a finite insertion loss and this inevitably

* The term dBm is often used to express the level of a signal with respect to a milliwatt. Thus +30 dBm means a signal level 30 dBs above 1 milliwatt, i.e. 1 watt and −30 dBm is similarly 1 microwatt.

brings about a degradation in the noise performance of the device. The travelling wave maser does not suffer from these defects and its development very shortly following that of the cavity maser resulted in cavity masers being employed in a relatively small number of systems. The principal interest of the cavity devices is that the validity of the maser principle and the practical feasibility of the maser was first demonstrated with cavities. In the next section we shall be concerned with the travelling wave maser which is the form of the device which has found real practical application.

3.3. Travelling Wave Masers

In the introduction to their paper on the cavity maser McWhorter and Meyer make the comment "To achieve the large bandwidths inherent in the widths of the paramagnetic resonance line without sacrificing gain a very low Q structure, e.g. a slow wave structure containing a much larger volume of the paramagnetic salt would have been necessary". The device which McWhorter and Meyer envisaged at that time was the Travelling Wave Maser—first suggested by H. Motz (1957).

As we have already discussed, if maser devices giving useful gain are to be obtained it is necessary that for a given signal power input the amplitude of the r.f. magnetic field should be made as large as possible. The use of the resonant cavity provides one means of obtaining a large r.f. field amplitude but an alternative method is to make use of a slow wave structure in which the velocity of propagation of electromagnetic energy is substantially less than the free space velocity. Field concentration occurs in such a structure as a consequence of the law relating the stored energy (W_s) per unit length of the structure to the power (P) in the wave travelling in the structure, i.e.

$$P = W_s v_g \tag{3.29}$$

v_g being the group velocity which in a lossless structure is equal to the energy velocity (see Watkins, 1958). This equality is approximately valid in practical situations where the loss is finite. If W_s is made large by velocity reduction then so too is the r.f. field. A maser constructed from a propagating structure uniformly loaded with maser material is a travelling wave maser and it possesses several important advantages over cavity devices. In the first place the r.f. energy emitted from each element of the maser material leads to an increase in energy flowing in the structure and, because the emitted radiation is in phase with the stimulating field, this leads to an increase in the rate of emission from succeeding elements but does not react back on the original element. Regeneration and oscillation are thus possible only as a result of reflections in the external circuitry. This is an important advantage over the cavity device; other advantages are, as we shall see, that the bandwidth varies only slowly with gain and is primarily limited by the paramagnetic resonance line width of the sample. Thirdly, it is possible to include non-reciprocal elements within the maser structure which are such as to make it completely stable even under high gain conditions and thus the need for external non-reciprocal elements is avoided. The non-reciprocal elements usually take the form of pieces of ferrite material suitably shaped so that resonant interaction occurs with the r.f. fields at the signal frequency for the same applied field as is necessary to give the required level splittings in the maser material. These elements are positioned in the propagating structure at points where the r.f. field is circularly polarised, the sense of polarisation being such that resonant interaction only occurs between the ferrite element and backward travelling fields. For example, if the T.W.M. consisted of a rectangular waveguide loaded with maser material (not usually a practical configuration) the isolators should be positioned as indicated in Fig. 19 where for an H_{10} mode

$$x = \frac{a}{\pi}\tan^{-1}\frac{\pi}{\beta a}; \quad \beta = \left[\left(\frac{\omega}{c}\right)^2 - \left(\frac{\pi}{a}\right)^2\right]^{\frac{1}{2}} \tag{3.30}$$

It is true that the non-reciprocal elements within the actual maser structure inevitably introduce a certain amount of forward loss but, partly because of the low temperature of the elements contri-

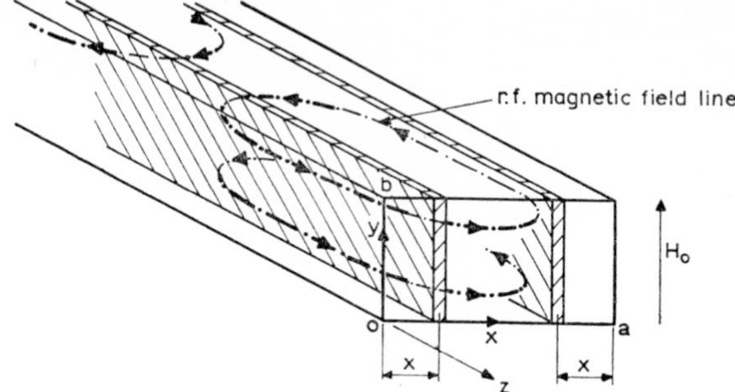

FIG. 19. Circular polarisation of r.f. magnetic field in a rectangular waveguide propagating an H_{10} mode. Circular polarisation occurs at the points shown where

$$x = \frac{a}{\pi} \tan^{-1} \frac{\pi}{\beta a}.$$

buting the loss the noise performance of the device is not affected significantly in consequence.

The characteristics of an amplifier which are of prime interest are gain, bandwidth (and the relation between them), noise figure and dynamic range and, before examining the pioneering Bell Laboratories paper on the T.W.M., it will be useful to consider the performance of the device in a manner consistent with our earlier analysis of the reflection cavity maser. Let us first consider the gain.

3.3.1. *The gain of the travelling wave maser*

To calculate the gain of the T.W.M. we define a magnetic Q factor Q_m and intrinsic Q, Q_0 in essentially the same way as we did for the cavity device but with the difference that whereas in the case of the cavity the quantities were defined in terms of the *total* energies stored, dissipated and emitted, for the T.W.M. the quantities will be defined per unit length of the structure. Considering an element of the maser of length dz the increment in power level in the element is

$$dP = (P_m - P_a)\, dz \tag{3.31}$$

(P_m and P_a being respectively the powers emitted by the maser material per unit length and that absorbed in the structure per unit length). Recalling our definitions of the Q factors

$$dP = 2\pi \, \nu_s W_s \left(\frac{1}{Q_m} - \frac{1}{Q_0}\right) dz$$

$$= 2\pi \, \nu_s \frac{P}{v_g} \left(\frac{1}{Q_m} - \frac{1}{Q_0}\right) dz \qquad (3.32)$$

(since $P = W_s v_g$).

Integrating this expression we find that the net gain (in decibels) of a T.W.M. of length L is

$$G_0 = 20\pi \log_{10} e \, \frac{\nu_s L}{v_g} \left(\frac{1}{Q_m} - \frac{1}{Q_0}\right)$$

$$\therefore G_0 = 27 \cdot 3 \left(\frac{c}{v_g}\right)\left(\frac{L}{\lambda}\right)\left(\frac{1}{Q_m} - \frac{1}{Q_0}\right);$$

λ = free space wavelength at the signal frequency. (3.33)

In practical masers, as we shall see (Section 3.4), values of Q_m of about 50 can be obtained at liquid helium temperatures and thus, if we are to obtain an electronic gain in excess of 50 dB in a device which is approximately one free space wavelength long (a practical length generally), it is necessary that

$$\frac{c}{v_g} \sim 100$$

For this reason slow wave structures are employed in T.W.M. and a suitable structure should provide a slowing factor (c/v_g) of about 100, contain regions where the r.f. magnetic fields are circularly polarised to facilitate the provision of reverse isolation and it should also permit the propagation of the pump frequency although not necessarily in a slow mode.

Structures which are suitable are those consisting of an array of parallel conductors within a rectangular waveguide, Fig. 4. The signal propagates in a slow mode associated with the array, the r.f. magnetic field being substantially circularly polarised with opposite senses on the two sides of the array. The pump propagates in a waveguide mode not greatly perturbed by the presence of the array which lies in a magnetic equipotential plane of that mode. The structure question is discussed at length in the paper by De Grasse *et al.* (1959) and it is interesting to note that the comb structure which they adopted has been employed almost exclusively in practical T.W. masers.

Although this derivation of the gain of the T.W.M. is commonly used and quoted it contains an implicit assumption which should not pass without comment. It is assumed that *all* the power emitted by the element travels in the forward direction and it is by no means self evident that this is the case. This point may be clarified to some extent if we regard each element of the maser material as a field source and remember that the fields associated with the stimulated emission are in phase with the stimulating field. If we consider propagation in a given direction on the structure, it is easy to see that, although each element of the maser material is a source of field increments of equal magnitude in the forward and backward directions, it is only the *forward* travelling field increments which add in phase and reinforce the forward travelling wave resulting in a growing wave. The backward-travelling increments interfere destructively but since we have a growing forward wave and the field increments are dependent on length the interference may not be complete. However, the residual backward-travelling wave will be very much smaller in amplitude than the growing forward wave. Thus since

$$dP \propto HdH$$

the major part of the power emitted by the element of the maser material flows in the forward direction, as we have assumed.

3.3.2. *The bandwidth of the travelling wave maser*

The instantaneous bandwidth of the travelling wave maser is primarily determined by the width and shape of paramagnetic resonance line employed since, as we will show in Section 3.4, the line shape function $g(\nu)$ (Section 2.4.2) occurs linearly in the expression for $1/Q_m$. It is frequently supposed, e.g. De Grasse *et al.* (1959), that the line shape is Lorentzian in which case $g(\nu)$ is given by equation (2.6), i.e.

$$g(\nu) = \frac{2T_2}{1+4\pi^2(\nu-\nu_0)^2T_2^2}$$

It follows then that the instantaneous bandwidth to 3 dB points (half power) of a T.W.M., giving an electronic gain G_e dB,* is

$$B = \frac{1}{\pi T_2}\sqrt{\left(\frac{3}{G_e-3}\right)} \qquad (3.34)$$

In fact in most maser materials (ruby in particular) the line shape is often much more nearly Gaussian than Lorentzian (Section 2.4.2). As it happens the two line shapes differ substantially only in the "wings" of the lines and in maser practice; as far as bandwidth is concerned, the distinction between Lorentzian and Gaussian line shapes is only of significance when the question of increasing the natural bandwidth of the maser by field staggering techniques is under consideration (see, for example, Walling and Smith, 1963).

3.3.3. *The noise temperature of the travelling wave maser*

The factors which determine the noise performance of the T.W.M. are identical with those in the case of the cavity maser, viz. loss occurring in the input waveguides, loss in the structure

* When we refer to *electronic* gain we mean the gain which the maser would give in the absence of any losses, i.e.

$$G_e = 27{\cdot}3\left(\frac{c}{v_g}\right)\frac{L}{\lambda}\frac{1}{Q_m}$$

itself (including, of course, that contributed by the ferrite isolators) and spontaneous emission from the maser material itself. Although the calculation of the noise temperature of the T.W.M. differs in detail from that for the case of the cavity maser because of the different configuration the result is the same in the two cases (to the same order of approximation), i.e.

$$T_m = \frac{1}{1-\lambda} [\lambda T_\lambda + | T_s |] \tag{3.35}$$

where the symbols have the same significance as previously. The noise temperature of the T.W.M. will thus be very similar to that of a cavity maser operating under similar conditions.

3.3.4. *The dynamic range of the travelling wave maser*

As in the case of the cavity maser the dynamic range of the T.W.M. is limited on the low side by the internal noise level. For a T.W.M. with a bandwidth (B) of 25 MHz and an equivalent noise temperature (T_m) of 10°K the lowest detectable input signal is $kT_mB = 3\times10^{-15}$ watt (-115 dBm). On the high side, again as for the cavity device, the limit is set by the gain saturation which occurs as the population difference between the signal transition levels begins to be reduced by the stimulated transitions (equation (1.21)). By integrating equation (3.32) for the case where $1/Q_m$ is not independent of P but depends on it as indicated by (1.21) we find that the gain G (dB) at an input level P_i is given by

$$\frac{G}{G_0} - 1 + \frac{P_0 - P_i}{P_s L} = 0 \tag{3.36}$$

In this expression (valid for $Q_0 \gg Q_m$, i.e. a low loss structure) G_0 is the small signal gain in dB, P_0 is the output power and P_s is defined in equation (3.28) (V_c being the volume of crystal per unit length for the T.W.M.). If the gain of the maser is large, i.e. $P_0 \gg P_i$, then equation (3.36) may be rewritten as

$$\frac{G_0 - G}{G_0} = \frac{P_0}{P_s L} \tag{3.37}$$

This is a particularly simple and useful result. It must be modified somewhat for the case of finite insertion loss.

In Fig. 20 we plot a saturation characteristic calculated from this expression for a T.W.M. with $P_sL = -16$ dBm—this is a fairly typical figure for an S-band T.W.M. using ruby as the active material.

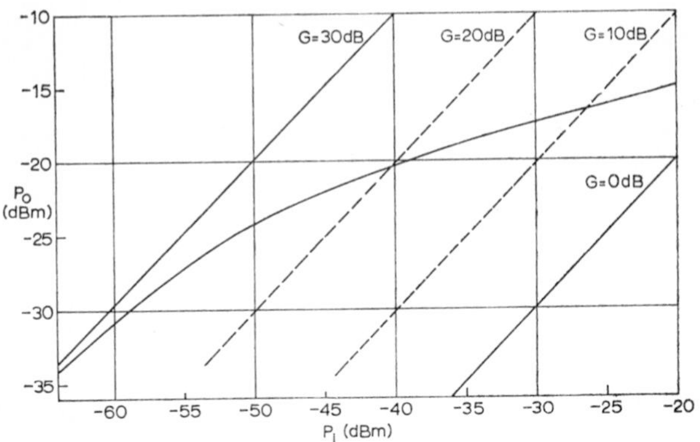

FIG. 20. Gain saturation in a travelling wave maser giving a small signal gain of 30 dB (calculated from equation (3.37) with $P_sL = -16$ dBm).

3.4. Determination of Q_m

The quantity which we have termed Q_m is of crucial importance in the design of solid state masers and it is necessary that we should derive an expression for Q_m in terms of accessible material and structure parameters. Our discussion follows that of Walling and Smith (1963).

Q_m has been defined as (equation (3.2))

$$Q_m = \frac{2\pi \nu_s W_s}{P_m}$$

and

$$P_m = \Delta n V_c h \nu_s W_{MM'}$$

It has already been shown (Section 2.8, equations (2.41) and (2.46)) that in general the probability of transitions between states $| M \rangle$ and $| M' \rangle$ being stimulated by an r.f. magnetic field $\boldsymbol{H}$ is

$$W_{MM'} = W_{M'M} = \tfrac{1}{4}\gamma^2 g(\nu) \,|\, (\boldsymbol{A}+i\boldsymbol{B}) \,.\, \boldsymbol{H} \,|^2 \tag{3.38}$$

In this expression γ is the gyromagnetic ratio $(2\pi g\beta/h)$ and $g(\nu)$ is a normalised line shape function (Section 2.4.2), $\boldsymbol{A}$ and $\boldsymbol{B}$ are matrix element vectors. In general the r.f. magnetic field will be elliptically polarised (linear and circular polarisations being particular cases) and thus

$$W_{MM'} = \tfrac{1}{4}\gamma^2 g(\nu) \,|\, (\boldsymbol{A}+i\boldsymbol{B}) \,.\, (\boldsymbol{H}+i\boldsymbol{H}') \,|^2 \tag{3.39}$$

$\boldsymbol{H}$ and $\boldsymbol{H}'$ being mutually perpendicular vectors. If we now consider an element of the maser material of volume dv the power emitted from it is

$$dP_m = \Delta nh\nu_s W_{MM'}\, dv \tag{3.40}$$

It is convenient to resolve the magnetic field into two circularly polarised components H_+ and H_- of opposite sense such that

$$H = H_+ + H_- \quad \text{and} \quad H' = H_+ - H_- \tag{3.41}$$

and then if $\boldsymbol{A}$ and $\boldsymbol{B}$ are coplanar with the r.f. field we find

$$dP_m = \Delta nh\nu_s \frac{\gamma^2}{4} g(\nu) \,[H_+^2(A+B)^2 + H_-^2(A-B)^2$$
$$+ 2H_+ \, H_- \cos 2\theta (A^2 - B^2)] \, dv$$

θ is the angle between $\boldsymbol{A}$ and $\boldsymbol{H}$ (the larger field component).

Integrating this expression over the volume of the crystal V_c

$$P_m = \Delta nh\nu_s \frac{\gamma^2}{4} g(\nu) \,[(A+B)^2 \int_{V_c} H_+^2 dv + (A-B)^2 \int_{V_c} H_-^2 \, dv$$
$$+ 2(A^2 - B^2) \int_{V_c} \cos 2\theta \, H_+ \, H_- \, dv]$$

In a few simple cases these integrals may be evaluated explicitly—

generally we have to resort to an approximate procedure. We define a reciprocity factor

$$R = \int_{V_c} H_-^2 \, dv \Big/ \int_{V_c} H_+^2 \, dv \tag{3.42}$$

whence to a good approximation

$$P_m = \Delta n h \nu_s \frac{\gamma^2}{4} g(\nu) \int_{V_c} H_+^2 \, dv [(A+B)^2 + R(A-B)^2 + 2 \cos 2\theta (A^2 - B^2) R^{\frac{1}{2}}]$$

It then follows that

$$\frac{1}{Q_m} = \frac{\Delta n h \gamma^2 g(\nu)}{2(1+R)\mu} \eta [(A+B)^2 + R(A-B)^2 + 2 \cos 2\theta (A^2 - B^2) R^{\frac{1}{2}}]\dagger \tag{3.43}$$

where

$$\eta = \int_{V_c} \mathbf{H} \, . \, \mathbf{H}^* \, dv \Big/ \int_{V} \mathbf{H} \, . \, \mathbf{H}^* \, dv \tag{3.44}$$

η is the ratio of the magnetic energy stored in the crystal to that stored in the whole structure; it is referred to as the filling factor.

It is useful to consider some particular cases to which (3.43) may be applied:

(i) if the r.f. field is circularly polarised $R = 0$ or ∞ depending on the sense of polarisation (3.42),

$$\therefore \frac{1}{Q_m} = \Delta n h \gamma^2 g(\nu) \eta \frac{(A \pm B)^2}{2};$$

(ii) for linear polarisation $R = 1$,

$$\therefore \frac{1}{Q_m} = \Delta n h \gamma^2 g(\nu) \frac{\eta}{2} [A^2 + B^2 + \cos 2\theta (A^2 - B^2)],$$

† Gaussian units are employed in this expression.

if the linearly polarised r.f. is parallel to A : $\theta = 0$ and

$$\frac{1}{Q_m} = \Delta n h \gamma^2 g(\nu) \eta A^2$$

if it is parallel to B : $\theta = 90°$ and

$$\frac{1}{Q_m} = \Delta n h \gamma^2 g(\nu) \eta B^2.$$

It also follows from the expression for Q_m that the maximum interaction (minimum Q_m) is obtained when $\theta = 0$ and

$$R^{\frac{1}{2}} = \frac{A-B}{A+B}$$

i.e. the r.f. magnetic field and the matrix elements have the same "polarisation", in this case

$$\frac{1}{Q_m} = \Delta n h \gamma^2 g(\nu) \eta (A^2 + B^2)$$

The results which we have derived for Q_m apply equally to the case of the cavity device where the fields are generally linearly polarised.

It will be useful to put numbers into these expressions remembering that we are using the Gaussian system. For the case where the active material is ruby (Cr^{3+} in Al_2O_3) and the signal transition at 4200 MHz is between levels 1 and 2, the magnetic field being applied at right angles to the three-fold axis of symmetry, the energy levels being then as indicated in Fig. 21, we find from Chang and Siegman's data (Chang and Siegman, 1958b) that

$\boldsymbol{A} = 1{\cdot}16$ and $\boldsymbol{B} = 0{\cdot}56$

also

$\gamma^2 = 3{\cdot}09 \times 10^{14}$ gauss^{-2} sec^{-2},

$h = 6{\cdot}62 \times 10^{-27}$ erg sec,

$g(\nu) \simeq 2T_2 = 10^{-8}$ sec for $0{\cdot}05\%$ Cr : Al ruby,

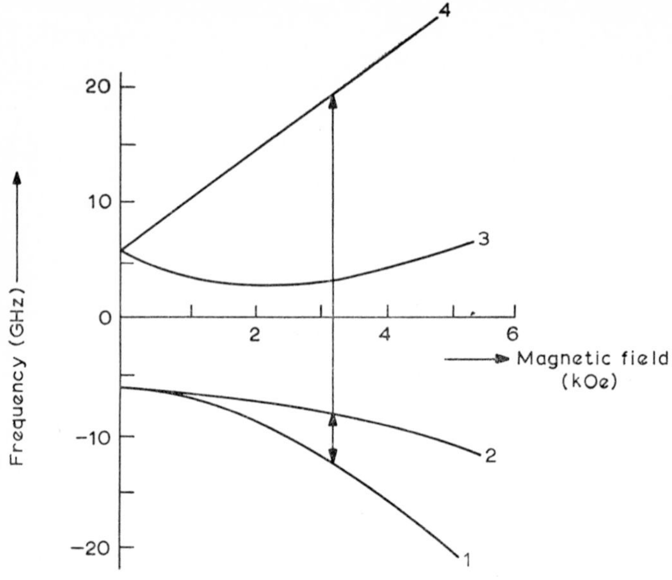

FIG. 21. Splitting of the ground state levels of ruby $\theta = 90°$.
$\nu_s(1\text{–}2) = 4170$ MHz,
$\nu_p(1\text{–}4) = 30{,}190$ MHz.

$$\Delta n = \frac{NIh\nu_s}{4kT};$$

N = total number of active ions per cc $= 2{\cdot}35 \times 10^{19}$ for $0{\cdot}05\%$ ruby,

I = inversion obtained by pumping ~ 3 in this case if the pump is applied between levels 1 and 4;

$$\therefore \Delta n \doteqdot \frac{4{\cdot}4}{T} \times 10^{18}\ \text{cm}^{-3};$$

$$\therefore \frac{1}{Q_m} = \frac{9 \times 10^{-2}}{T}\ \eta\ M,$$

where

$M = A^2+B^2 = 1{\cdot}66$ optimum case,

$M = \dfrac{(A \pm B)^2}{2} =$ 1·5 or 0·18 for circular polarisation depending on the sense,

$M = A^2 = 1{\cdot}34$ for linear polarisation $H \parallel A$,

$M = B^2 = 0{\cdot}31$ for linear polarisation $H \parallel B$.

Thus for this case with a filling factor (η) of 0·3 (a practical figure) and an operating temperature of 4·0°K we find with a circularly polarised r.f. field

$$\frac{1}{Q_m} \simeq 10^{-2}; \quad Q_m = 100$$

The inversion obtained with a particular pumping scheme depends both on concentration and temperature and the figure of 3 which we have chosen for I is not an optimum; it is, however, adequate to show the sort of values of Q_m which are achievable in practice.

CHAPTER 4

The Application of Solid State Masers

The outstanding characteristic of the Solid State Maser is its extremely low equivalent noise temperature and this has led to its employment as the first-stage amplifier in satellite communication systems. This application is discussed in the extracts from the paper by Walling and Smith which follows. For use in satellite communication systems having commercial viability it is desirable that the intrinsic bandwidth of the T.W.M. (such as that described by Walling and Smith) should be increased. As we have already mentioned the bandwidth of the maser can be increased if the magnetic field acting on the maser crystal is varied along the length of the device. Naturally this results in a reduction of the gain of the maser and in Fig. 22, taken from the Walling and Smith paper, bandwidth is plotted as a function of electronic gain for several different types of field variation for the case of a maser giving a uniform field gain of 50 dB with 16 MHz bandwidth to 3 dB points. From this figure (in which a Lorentzian line shape is assumed) it is apparent that bandwidths in excess of 50 MHz can be realised without paying a prohibitive price in terms of gain reduction.

Although the main attribute of the T.W.M. which led to its employment in satellite communication systems is its exceedingly low noise temperature the device has other characteristics which make it outstandingly good for communications purposes. These are its linearity and intermodulation characteristics. Earlier we discussed the behaviour of the T.W.M. under large signal conditions and noted that gain saturation occurred in the device as a consequence of the population redistribution which occurs under

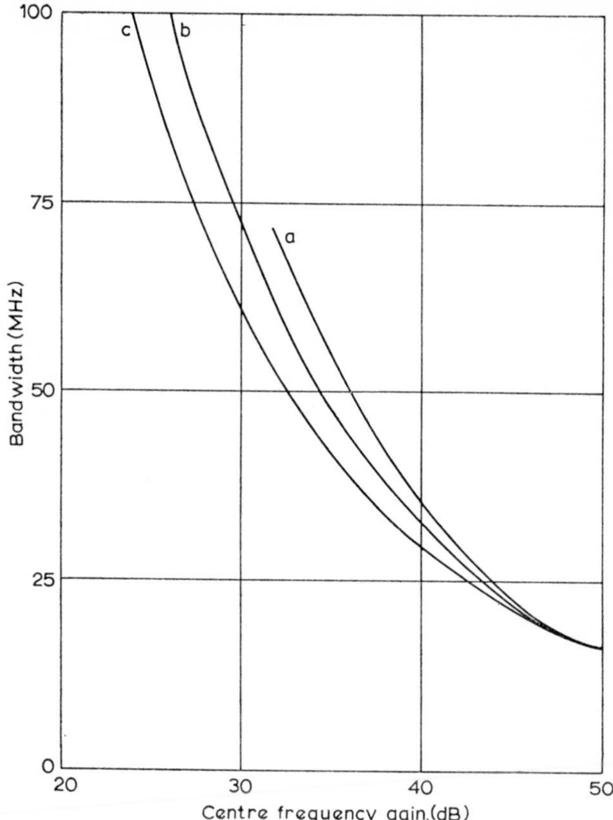

FIG. 22. Bandwidth increase in a T.W.M. by field staggering techniques (*a*) Single step field variation. (*b*) Sinusoidal field variation. (*c*) Linear field variation. These curves are calculated on the assumption of a Lorentzian line shape.

the combined effects of the signal, pump and spin lattice interaction processes. The level population changes, however, do not occur instantaneously but in a time comparable with the spin–lattice relaxation time and thus under pulsed conditions gain saturation occurs at peak power levels which exceed the C.W. values in direct proportion to the duty cycle (the pulse length being assumed to be less than the spin–lattice relaxation time). Because of the difference in saturation behaviour with slow and fast changes in input power level the T.W.M. behaves as a linear amplifier even when operating with compressed gain. This situation is illustrated in Fig. 23. Suppose we have a maser operating at point *A* where

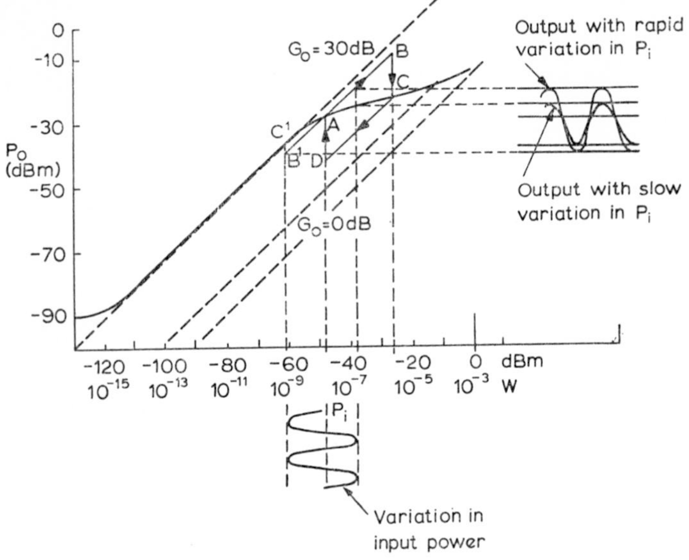

FIG. 23. Dynamic characteristic of a T.W.M. giving 30 dB small signal net gain.

there is slight gain saturation, if the input level is now raised rapidly the gain is initially unchanged (because the populations of the levels have not had time to readjust themselves to the new situation) and the output is given by point *B*. However, if the input power level is maintained at *Pi′* the gain will decay slowly to that given by point *C*; a rapid reduction in input power to its original value takes us to point *D* where the gain is the same as that at point *C*. From point *D* the gain will recover to the equilibrium of point *A*. The recovery, of course, is again a relatively slow process. Thus gain compression, although it exists, in the T.W.M. does not lead to non-linearity. To emphasise the point, unless the changes in amplitude of the input signal occur in a time comparable with or greater than the spin–lattice relaxation time the output power will be linearly related to input following the gain characteristic *B'AB* (Fig. 23). If, on the other hand, the changes in amplitude of the input signal occur slowly then the output power will not be linearly related to the input following the static gain characteristic *C'AC*. The output power will thus be distorted and contain harmonics. Since spin–lattice relaxation times in ruby are of the order of 100 msec distortion due to this cause will only be observed at signal modulation frequencies of about 10 Hz—much lower than will occur in practice. This aspect of the maser's behaviour is unique. Intermodulation effects do occur in the T.W.M. but they are not associated with gain compression as in other amplifiers. They arise only when there is an appreciable degree of coherence in the motion of the spin system which is generally not established unless very high r.f. magnetic fields are applied, overcoming the strong tendency of spin–spin interactions to destroy such coherence. This question has been discussed by Schulz-du Bois (1964) who shows that for a typical T.W.M. the second harmonic power generated is (in dBM)

$$P(2\nu_1) = 2P(\nu_1) - 103{\cdot}5 \qquad (4.1)$$

and the second-order intermodulation power

$$P(2\nu_1 - \nu_2) = 2P(\nu_1) + P(\nu_2) - 92 \qquad (4.2)$$

Under C.W. conditions with an output power at $(P(\nu_1))$ of $-$ 30 dBm (typical) the second harmonic power generated will be at a level of $-163{\cdot}5$ dBm and the intermodulation product $(2\nu_1 - \nu_2)$ will be at -182 dBm (assuming power levels at the frequencies ν_1 and ν_2 to be -30 dBm). Both these power levels are far below the detectable limit of the maser (-90 dBm).

The reader will recognise that the origins of second harmonic generation and intermodulation in the T.W.M. (or masers generally) are anologous to the process occurring in ferrite devices but, of course, the effects are much less because of the very much weaker spin–spin interaction effects in the paramagnetic material of the maser.

To all intents and purposes then the T.W.M. behaves as a linear amplifier even when operating with compressed gain and it is free from intermodulation effects. These attributes, together with its uniquely low noise temperature, make it an ideal amplifier for use in satellite communication systems. It is indeed no exaggeration to say that the early satellite communication systems were successful as a result of the employment of T.W.M. in the ground station receivers (De Grasse *et al.*, 1961—Echo; Tabor and Sibilic, 1963; Walling and Smith, 1963).

Other areas of application of masers are in the tracking of deep space probes and T.M.W. are used in conjunction with closed-cycle refrigerators to provide the low-temperature environment in the N.A.S.A. (National Aeronautics and Space Administration) Deep Space Instrumentation Facility which is comprised of tracking stations separated in longitude by about 120°; there are six stations in the system. In this application overall system noise temperature rather than bandwidth is of prime importance. The use of T.W.M.s operating at a frequency of 2290 MHz has substantially increased the data rate capability of the DSIF receivers and made possible the transmission and reception of television pictures of the surface of Mars via the *Mariner* satellite.

Masers, both of the cavity type and the T.W.M., have also found application in the radio astronomy field and these applications are discussed by Jelley and Cooper (1961) and by Smith *et al.* (1966).

Conclusion

Little over twelve years has elapsed since the invention of the solid state maser and in that time a very large amount of research and development effort has been devoted to the device and its concomitant material problems. This has resulted in the rapid realisation of sophisticated low noise maser systems which have been employed with conspicuous success in the pioneering experiments in satellite communications (Projects, Echo, Telstar and Relay) and indeed are still employed in such systems. It is nevertheless true that despite the large amount of work which has been carried out on maser materials and techniques, there has been no indication that low noise masers having bandwidths adequate for communication purposes can be realised at temperatures higher than that of liquid helium (4·2°K) and even at helium temperatures it is clear that the attainment of instantaneous bandwidths in excess of 100 MHz together with acceptable gain ($>$ 20 dB) is a matter of considerable difficulty.

These limitations of the maser, together with the very rapid developments which have taken place in the area of diode parametric amplifiers, have resulted in a virtual cessation of development in the maser field. The maser story is thus not one of continuing development but is, to all intents and purposes, now at an end. There can, however, be few instances in the history of applied science where a somewhat esoteric branch of pure science has been so rapidly exploited in the realisation of devices meeting, uniquely, an important practical necessity and the solid state maser will undoubtedly occupy an honourable place in scientific history.

References

ALLEN (1969) *Essentials of Lasers*, Pergamon, Oxford.

ARTMAN, J. O., BLOEMBERGEN, N. and SHAPIRO, S. (1958) *Phys. Rev.* **109,** 1392.

ASSENHEIM, H. M. (1966) *Introduction to Electron Spin Resonance*, Hilger & Watts, London.

BASOV, N. G. and PROKHOROV, A. M. (1955) *J.E T.P.* (*U.S.S.R.*) **28,** 249.

BLOCH, F. (1946) *Phys. Rev.* **70,** 460.

BLOEMBERGEN, N. (1956) *Phys. Rev.* **104,** 324.

BLOEMBERGEN, N., SHAPIRO, S., PERSHAN, P. S. and ARTMAN, J. O. (1959) *Phys. Rev.* **114,** 445.

BRICE, J. C. (1965) *The Growth of Crystals from the Melt*, North Holland Publ. Co.

BUTCHER, P. N. (1958) *Proc. I.E.E.* **105,** 684.

CHANG, W. S. and SIEGMAN, A. E. (1958a) Stanford Technical Report 156–1.

CHANG, W. S. and SIEGMAN, A. E. (1958b) Stanford Technical Report 156–2. The results obtained in this report are quoted by WEBER (1959).

COMBRISSON, J., HONIG, A. and TOWNES, C. H. (1956) *Comptes Rendus* **242,** 2451.

DE GRASSE, R. W., SCHULZ-DU BOIS, E. O. and SCOVIL, H. E. D. (1959) *Bell Syst. Tech. J.* **38,** 305.

DE GRASSE, R. W., KOSTELNICK, J. J. and SCOVIL, H. E. D. (1961), *Bell Syst. Tech. J.* **40,** 1117.

DEVOR, D. P., D'HAENENS, I. J. and ASAWA, C. K. (1962), *Phys. Rev. Letters* **8,** 432.

DIRAC, P. A. M. (1927) *Proc. Roy. Soc.* A **114,** 243, 710.

EINSTEIN, A. (1917) *Phys. Z.* **18,** 121.

FLETCHER, R. (1952) *Proc. I.R.E.* **40,** 951.

FONER, S., MOMO, L. R., THAXTON, J. B., HELLER, G. S. and WHITE, R. M. (1961) *Advances in Quantum Electronics*, p. 553 (Ed. J. R. SINGER), Columbia, New York.

GEUSIC, J. E. (1956) *Phys. Rev.* **102,** 1252.

GORDON, J. P., ZEIGER, H. J. and TOWNES, C. H. (1954) *Phys. Rev.* **95,** 282.

GORDON, J. P., ZEIGER, H. J. and TOWNES, C. H. (1955) *Phys. Rev.* **99,** 1264.

HINDMARSH, W. R. (1967) *Atomic Spectra*, Pergamon, Oxford.

HUGHES, W. E. (1962) *Proc. I.R.E.* **50,** 1691.

INGRAM, D. J. E. (1967) *Spectroscopy at Radio and Microwave Frequencies*, 2nd Ed., Butterworths, London.

JELLEY, J. V. and COOPER, B. F. L. (1961) *Rev. Sci. Inst.* **32,** 166.

KITTEL, C. (1958) *Elementary Statistical Physics*, John Wiley, New York.

KYHL, R. L., MACFARLANE, R. A. and STRANDBERG, M. W. P. (1962), *Proc. I.R.E.* **50,** 1608.

LOW, W. (1960) *Paramagnetic Resonance in Solids*, Supplement 2, Solid State Physics, Academic Press, New York.

MCLAUGHLAN, S. D. (1963) *R.R.E. Memo*. 2018.

MCWHORTER, A. L. and MEYER, J. W. (1958) *Phys. Rev.* **109,** 312.

MAIMAN, T. H. (1960) *Advances in Quantum Electronics*, vol. II, p. 324 (Ed. TOWNES, C. H.), Columbia, New York.

MAKHOV, G., KIKUCHI, C., LAMBE, J. and TERHUNE, R. W. (1958) *Phys. Rev.* **109,** 1399.

MOTZ, H. (1957) *J. Elect. and Control* **2,** 571.

NAGY, A. W. and FRIEDMAN, G. E. (1962) *Proc. I.R.E.* **50,** 2504.

ORTON, J. W. (1968) *Electron Paramagnetic Resonance*, Iliffe, London.

ORTON, J. W., FRUIN, A. S. and WALLING, J. C. (1966) *Proc. Phys. Soc* **87,** 703.

PACE, J. H., SAMPSON, D. F. and THORP, J. S. (1960) *Proc. Phys. Soc.* **76**, 697.

PAKE, G. E. (1962) *Paramagnetic Resonance*, Benjamin, New York.

PAULING, L. and WILSON, E. B. (1935) *Introduction to Quantum Mechanics*, Chap. 6, McGraw-Hill, New York.

POOLE, C. P. (1967) *Electron Spin Resonance*, Interscience, New York.

PURCELL, E. M. and POUND, R. V. (1951) *Phys. Rev.* **81,** 279.

ROBERTS, J. K. (1940) *Heat and Thermodynamics*, p. 385, 3rd Ed., Blackie, London.

SABISKI, E. S. and ANDERSON, C. H. (1966) *Appl. Phys. Letters* **8,** 298.

SCHULZ-DU BOIS, E. O. (1959) *Bell. Syst. Tech. J.* **38,** 271.

SCHULZ-DU BOIS, E. O. (1964) *Proc. I.E.E.E.* **52,** 644.

SCOVIL, H. E. D., FEHER, G. and SEIDEL, H. (1957) *Phys. Rev.* **105,** 762.

SLATER, J. C. (1960) *Quantum Theory of Atomic Structure*, Vol. 1, Chap. 11, McGraw-Hill, New York.

SMITH, F. W., BOOTH, P. L. and HENTLEY, E. L. (1966) *Philips Technical Review* **27,** 313.

STERN and GERHLACH (1922) See RICHTMEYER, F. K., KENNARD, E. H. and LAURITSEN, T. *Introduction to Modern Physics*, 5th Ed., p. 298, McGraw-Hill, New York.

TABOR, W. J. and SIBILIC, J. T. (1963) *B.S.T.J.* **52,** 1863.

TER HAAR, D. (1967) *The Old Quantum Theory*, Pergamon, Oxford.

TINKHAM, M. (1964) *Group Theory and Quantum Mechanics*, McGraw-Hill, New York.

VAN VLECK, J. H. (1940) *Phys. Rev.* **57,** 426.

WALLING, J. C. and SMITH, F. W. (1963) *Philips Technical Review* **25,** 289.

WATKINS, D. (1958) *Topics in Electromagnetic Theory*, Wiley, New York.

WEBER, J. (1953) *Trans. I.R.E.* PGED–3.

WEBER, J. (1959) *Rev. Mod. Phys.* **31,** 681.

ZVEREV, G. M., PROKHOROV, A. M. and SHEVCHENKO, A. K. (1964) *Advances in Quantum Electronics*, vol. III, p. 963, Columbia, New York.

PART II

Comments on Paper 1

THE experiments performed are on the nuclear magnetic resonance of Li^7 nuclei. This nucleus has a nuclear spin of 3/2 with a moment of $+3{\cdot}25$ nuclear magnetons. Two important points had been established by Purcell and Pound in a previous experiment. The first was that the relaxation time of the Li^7 nuclei at 6376 gauss was 300 sec and that as the field was reduced this value decreased falling to 15 sec at zero field. The second point was that at this field a resonance was observed at 50 kHz. The relaxation time was so long that the experiment described in the paper could be easily carried out. This consists basically of allowing the nuclear system to come to equilibrium at a field of 6376 gauss; the specimen is then quickly moved to a subsidiary magnet in which the field is rapidly reversed and the sample returned to the main magnet in which the population is sampled. In the small magnet the spins are subjected to a steady field of 100 gauss parallel to the main field. The spin system may be regarded as precessing about this field at the Larmor frequency, which in this case is of the order of 3×10^5 radians/sec as shown by the nuclear resonance frequency of 50 kHz.

If the magnetic field is slowly altered in direction at an angular frequency much less than the period of the Larmor precession, then the precessing spins can follow the motion of the field. By discharging a condenser through a coil surrounding the small magnet, the field is first reversed quickly in a time shorter than the precessional frequency of the spins. In this case the spin system cannot follow the rapid change and thus continues to precess about the old field direction with the result that the relative energies of the precessional motions become reversed. The magnetic field is now allowed to revert slowly to the original direction. This time the change is slow enough to allow the spins to precess around the field as it changes back to the original direction with a time constant of 1 msec. This is the adiabatic process referred to in the last paragraph. The result of this operation is that the spin system is now reversed and the upper levels have a greater population than the lower. Such a system is not in thermal equilibrium, of course, and thermalising process would eventually restore the populations to those given by the Boltzmann distribution.

The reversal of population resulting from these manipulations is clearly shown in Fig. 1 of the paper. The traces represent consecutive traverses through the resonance condition, the width of the trace showing the line width of the nuclear resonance and the height representing the spin population difference. The first trace on the left represents the situation when equilibrium has been obtained at 6376 gauss. The second trace shows the result after reversal of the population produced by the field reversal. The amplitude of this trace is only 0·75 of the first due to the decay processes taking place during the period of manipulation at low field. The reversed population decays back through zero to the equilibrium value with the high field time constant of 300 sec.

The concept of spin temperature is introduced to describe the populations of the levels in terms of the Boltzmann equation:

$$\frac{n_1}{n_2} = \exp -\Delta/kT$$

where Δ is the separation of the levels and level 1 is at higher energy than level 2, n_1 and n_2 being the populations. If $n_1 = n_2$ then this can be described by putting $T = \infty$, and if $n_1 > n_2$ then T becomes negative. Large population differences therefore result from a large level separation, or a very low temperature. Thus the situation at 6376 gauss which is described by $T = 300°K$ becomes described by $T \simeq -1°K$ when inverted at zero field where $\Delta = 50$ kHz and is only $T \simeq -350°K$ on return to the main field due to the losses in the cycling process.

It is interesting to note that Purcell and Pound mention that such a system is capable of inducing radiation because it is effectively very hot. Thus if the sample had been large enough to overcome the losses in the circuits used to observe the resonance, amplification and oscillation may have taken place. These losses are quite high for the oscillatory circuits used for nuclear magnetic resonance, whereas in the electron spin, resonance case microwave circuits having very low losses are used and the task of producing oscillation made somewhat easier.

PAPER 1*

A Nuclear Spin System at Negative Temperature

E. M. PURCELL and R. V. POUND

Department of Physics, Harvard University, Cambridge, Massachusetts

November 1, 1950

A NUMBER of special experiments have been performed with a crystal of LiF which, as reported previously,[1] had long relaxation times both in a strong field and in the earth's field. These experiments were designed to discover the conditions determining the sense of remagnetization by a strong field when the initially magnetized crystal was put for a brief interval in the earth's field.

At field strengths allowing the system to be described by its net magnetic moment and angular momentum, a sufficiently rapid reversal of the direction of the magnetic field should result in a magnetization opposed to the new sense of the field. The reversal must occur in such a way that the time spent below a minimum effective field is so small compared to the period of the Larmor precession that the system cannot follow the change adiabatically. The experiments in zero field reported above[2] showed a zero field resonance at about 50 kc and therefore the following experiment was tried.

The crystal, initially at equilibrium magnetization in the strong (6376 gauss) field, was quickly removed, through the earth's field, and placed inside a small solenoid, the axis of which was parallel

* *Phys. Rev.* **81**, 279–80 (1951).

(1) R. V. Pound, *Phys. Rev.* **81**, 156 (1951).

(2) N. F. Ramsey and R. V. Pound, *Phys. Rev.* **81**, 278 (1951).

to a field of about 100 gauss, provided by a small permanent magnet. A 2 μfd condenser, initially charged to 8 kv, was discharged through the coil, with 500 ohms in series, in such a sense that the field in the coil reversed to about -100 gauss, with a time constant of about 0.2 μsec and decayed back to the original field with a time constant of 1 msec. The crystal was quickly returned, through the earth's field, to the strong magnet and the Li^7 resonance sampled. The operation could be done in 2 to 3 sec. A reversed deflection was found and it decayed, through zero, to the

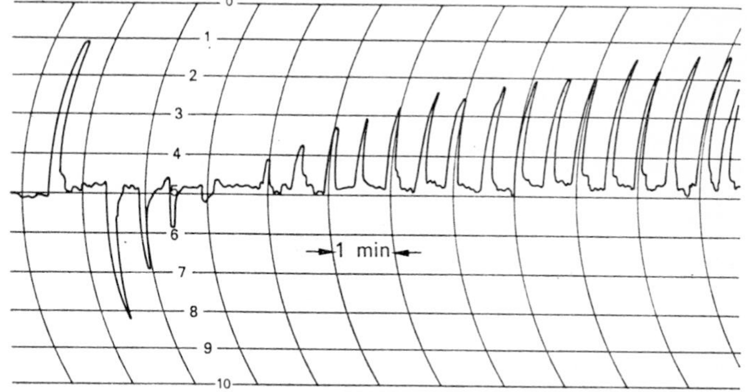

FIG. 1. A typical record of the reversed nuclear magnetization. On the left is a deflection characteristic of the normal state at equilibrium magnetization ($T \approx 300$°K), followed by the reversed deflection ($T \approx -350$°K), decaying ($T \rightarrow -\infty$) through zero deflection ($T = \infty$) to the initial equilibrium state.

equilibrium state with the characteristic 5-min time constant. A typical record is shown in Fig. 1.

The state of spin system just after this treatment is thought to be properly described by a negative spin temperature. The system loses internal energy as it gains entropy, and the reversed deflection corresponds to induced radiation. Statistically, the most probable distribution of systems over a *finite* number of equally spaced energy levels, holding the total energy constant, is the Boltzmann distribution with either positive or negative temperature determined by whether the average energy per system is smaller or larger, respectively, than the mid-energy of the available levels. The sudden reversal of the magnetic field produces the latter situation.

One needs yet to be convinced that a single temperature adequately describes the nuclear spin state. Bearing on this is the fact that the crystal passes through the earth's field after the inverted population is produced, on its way back to the main magnet. The retention of the reversed magnetization requires that the spin-only-state, in the earth's field, have an inverted population and be described by a suitably small (~ -1°K) negative temperature. Thus a very short time is required for the attainment of thermal equilibrium within the spin system itself (not the ordinary T_2, however).

A system in a negative temperature state is not cold, but very hot, giving up energy to any system at positive temperature put into contact with it. It decays to a normal state through infinite temperature.

This and related experiments indicate that the spin system is able to follow changes in even a small field adiabatically unless they occur in a time presumed to be less than about 20 μsec.

Comments on Paper 2

IT IS shown that a system with an inverted population is capable of emitting radiation. Two ways of obtaining the necessary reversal are considered, either the sudden reversal of a magnetic field or the reversal of an electric field. In either case the amplification will last only for a short interval. The author makes the ingenious suggestion that in the case of gases continuous operation could be obtained by a flow of the gas through the region of field reversal.

The propagation constant α referred to is the incremental power gain of the system. Thus $P_{out} = P_{in}\, e^{\alpha l}$ for active length l where P_{out} is the output power and P_{in} is the input. The constant α is given by:

$$\alpha = \frac{1}{P}\frac{dP}{dl} \tag{1}$$

In the case of a continuously inverted medium α can be calculated:

$$dP = \Delta N h\nu\, W_s$$

For thickness dl, ΔN is given by approximately:

$$\Delta N \simeq \frac{N_1}{kT} h\nu \,.\, dl$$

where N_1 is the number of ions in level 1 per cm before inversion.

For an electric dipole transition

$$W_s = \left(\frac{2\pi}{h}\right)^2 .\, E^2\, \mu^2\, \tau$$

where τ is the spin–spin relaxation time and μ is the electric dipole moment. Taking $P = 1/8\pi\ E^2c$ as input power, α is calculated from (1) as

$$\alpha \simeq \frac{32\pi^3 \,.\, \nu^2 \,.\, \mu^2}{ckT}$$

which is similar in form to the estimate made in the third equation of the gain under pulse conditions.

PAPER 2*

Amplification of Microwave Radiation by Substances not in Thermal Equilibrium

J. WEBER

Glenn L. Martin College of Engineering and Aeronautical Sciences, University of Maryland, College Park, Maryland

Introduction

This paper briefly discusses the possibility of obtaining coherent microwave radiation from crystals and gases. It will be shown that it is *possible* to obtain coherent microwave radiation by such methods, provided that a certain non-equilibrium energy distribution is first produced. Methods are discussed for producing such a distribution. The amount of amplification which can be produced by such methods is *very small* under ordinary circumstances and does not appear to be able to compete with other methods. The method may have certain special applications.

Many substances show a strong selective absorption of radiation at certain frequencies in the microwave region. This absorption comes about in the following way. Let us consider that there are two energy levels E_1 and E_2 where $E_1 < E_2$. Let N_1 be the number of oscillators in the state with energy E_1. Let N_2 be the number of oscillators in the state with energy E_2. If we have equilibrium

$$N_2 = N_1\, e^{-[(E_2-E_1)/kT]}$$

where k is Boltzmann's constant and T is the absolute temperature.

* *Trans. I.R.E.*, PGED–3 (1953).

Thus there will be, in equilibrium, always more oscillators in the lower energy state. Suppose we now apply radiation of frequency ν to the system where

$$\nu = \frac{E_2 - E_1}{h}$$

and h is Planck's constant. If the transition is allowed by the selection rules, the radiation will induce transitions for the oscillators with energy E_1 to the state with energy E_2. Also, the radiation will induce transitions downward (forced emission) of oscillators in the state E_2. It is well known from quantum theory that the transition probability *up* from E_1 to E_2 is the same as the transition probability *down* from E_2 to E_1; if we denote the transition probability from 1 to 2 as P_{12} and the transition probability from 2 to 1 as P_{21} then $P_{12} = P_{21}$. The rate of absorption of energy by oscillators in the lower state is $\text{Power}_{(\text{absorbed})} = P_{12}N_1\,(E_2 - E_1) = P_{12}N_1 h\nu$. The rate of emission by oscillators in the upper state is

$$\text{Power}_{(\text{emitted})} = P_{21}N_2\,(E_2 - E_1) = P_{21}N_2 h\nu$$

If we have equilibrium then, from the Maxwell–Boltzmann distribution law,

$$N_2 = N_1\, e^{-h\nu/kT} \approx N_1\left(1 - \frac{h\nu}{kT}\right)$$

Therefore we obtain

$$P_{\text{absorbed}} - P_{\text{emitted}} = \frac{P_{12}(h\nu)^2 N_1}{kT},$$

and this is a positive quantity. Thus under ordinary circumstances we get absorption of radiation (ordinary microwave spectroscopy) because the transition probability up is the same as the transition probability down, but since there are more oscillators in the lower states, we get a net absorption.

A Method of Obtaining Amplification

We could get amplification if somehow the number of oscillators in the upper states could be made greater than the number in the lower states. A method of doing this is suggested by Purcell's[(1)] negative temperature experiment.

Suppose we apply a magnetic field to an ensemble of molecules of a gas, or nuclei of a crystal lattice. If the particles have a magnetic dipole moment this will give a resultant polarization. The distribution in angle of the various moments at equilibrium is given by

$$e^{\mu H \cos\theta/kT}$$

where μ is the dipole moment, H is the magnetic field, θ is the angle between the dipole moment and the field.

It is known from the quantum theory that there is space quantization, only certain values of θ are allowed. Some of these are parallel to the field and some are anti-parallel. Suppose we have two states with values θ_1 and θ_2, then if we apply a radio-frequency field in the proper direction of frequency ν such that

$$h\nu = \mu H [\cos\theta_1 - \cos\theta_2],$$

and if the transition is allowed by the selection rules, we will get transitions from state one to state two. However, the higher energy states are less heavily populated, and the substance will absorb radiation. If now we suddenly reverse the magnetic field then it will take time to restore equilibrium (relaxation time). During this brief time, because the field has been reversed, we will have more oscillators in the upper state than in the lower state. We have an "upside down" Boltzmann distribution and during this time an incident electromagnetic wave will be amplified, and the resultant pulse of radiation will be coherent.

(1) E. M. Purcell and R. V. Pound, *Phys. Rev.* **81**, 279 (1951).

Symmetric Top Molecules in Electric Fields

The same thing can be done (much more easily) with polar symmetric top molecules in an electric field which is reversed. We then obtain coherent pulses of the amplified driving signal. We can also imagine a gas flowing continuously through a region in which an electric field reversal takes place. In the region of field reversal we would get amplification so that continuous rather than pulsed radiation would be emitted. It is essential that the Stark effect be linear in the applied field, not quadratic, otherwise field reversal will not yield the desired energy distribution.

For the linear Stark effect in symmetric top molecules the frequency which could be amplified is given approximately by

$$\nu = \frac{\mu E K}{J(J+1)h}$$

where ν is the frequency, μ is the permanent dipole moment, E is the electric field (Stark field), h is Planck's constant, J and K are rotational quantum numbers associated with the symmetrical top. The Stark field has to be perpendicular to the microwave electric field.

An electromagnetic wave will grow if propagated in a gas with a "negative temperature" energy distribution. The real part of the propagation constant would be roughly given by

$$\alpha = -\frac{8\pi^3 \nu^2 N_1 \mid \mu_1 \mid^2 \tau}{3c\, kT}$$

where ν is the frequency, N_1 is the number of molecules per unit volume in the originally lower state, $\mid \mu_1 \mid^2$ is the squared matrix element corresponding to the dipole moment of the transition, τ is the pulse duration, c is the velocity of light, k is Boltzmann's constant, T is the absolute temperature.

It should be emphasized that we will get gain only during a very short period before the molecular energy distribution begins to approach the equilibrium value.

Collisions tend to restore equilibrium. The applied microwave field tends to equalize the population of the two levels because transitions out of a state are proportional to the population of that state. Thus if one state is more heavily populated than another, transitions from that state occur faster and this tends to equalize the level population. This effect of the microwave field on level population is called saturation.[(2)]

Rough calculations have been made of the gain and maximum power to be expected in the simple case of a gas at low pressure. The Stark effect in the ammonia rotation spectrum (not inversion spectrum) was considered. It was found that for a pressure of 10^{-2} mm of Hg at $\nu = 30{,}000$ Mc/s, gains of the order of ·02 decibels per meter could be obtained and the saturation power was of the order of 2 milliwatts per 100 cc of gas. These figures could be enormously increased if a suitable transition (and relaxation time) could be found in either a solid or liquid.

(2) R. Karplus and J. Schwinger, *Phys. Rev.* **73,** 1020 (1948).

Comments on Paper 3

THIS paper describes the operation of a microwave oscillator at 23,870 MHz, the energy being obtained from an inverted population of ammonia molecules. The NH_3 molecule has the configuration of a pyramid, the base having the three hydrogen atoms at the corners with the nitrogen atom above or below the plane of the hydrogens. Because of the possibility of the nitrogen atom being on either side of this plane of atoms, the potential well in which the nitrogen moves will exhibit two minima separated by a hill. Such potential systems always give rise to symmetric anti-symmetric wave functions (Schiff, *Quantum Mechanics*, McGraw-Hill, vol. 1, sec. 3) and a doubling of each energy level of the system. In this case the lowest molecular levels are separated by 0·8 cm^{-1} (26·4 kMHz). Each of these levels is further split into other levels due to the various rotational states of the molecule and to the various combinations of the nuclear moments of the atoms present. The absorptions occurring in the region 20–26 kMHz are termed the ammonia inversion spectrum and consist of thirty main lines. Each of these lines can be designated by two quantum numbers J and K representing the total angular momentum and the projection of this on the molecular axis of symmetry. In all cases the lines with $J = K$ are the strongest, the line designated by $J = 3$, $K = 3$ or the 3,3 line at 23,870 MHz being the strongest of all. This line is shown in Fig. 2.

In addition to having different energies the two lowest states of the NH_3 molecule have different behaviour in an electric field, there being a difference in the electric dipole of the molecule in the two states. If the field has a gradient then those molecules in the high-energy state experience a force directed towards the region of high field; the opposite effect occurs for the low-energy state molecules.

In the experiment described here a beam of NH_3 molecules is passed through a system of electrodes which produces a cylindrical electrostatic field so arranged that the molecules in the low-energy state move outwards from the centre of the beam to the electrodes, while the molecules in the higher state are focused towards the centre of the electrode system. To achieve good separation, high voltages are required and the faces of the electrodes must be shaped to give high field gradients. For focusers of 22 in. in length spaced 0·08 in. apart 15 kV was used (these details were given in a later paper). The action of the focuser thus resulted in a beam of ammonia molecules all in the upper energy state which was allowed to flow into a large microwave cavity. This cavity could be tuned to the inversion frequency of 24 kMHz. In addition power was fed into the cavity from a frequency modulated klystron. In the absence of the beam the output from the cavity measured the losses present in the system. When the beam entered the cavity, transitions were induced by the radiation from the klystron as it swept through the inversion frequency and the molecules reverted to the lower energy state giving up energy to the cavity and causing the amplification shown in Fig. 2.

The power delivered to the cavity depends upon the number of molecules entering the cavity per second, the transition probabilities and the quantum given up in each transition. To cause oscillation the flow of molecules must be sufficiently great for this energy to exceed the cavity losses. With a cavity Q of 12,000 it was calculated that a flow of 10^{13} molecules/sec was required to produce oscillations.

Though this device is capable of continuous operation, the fact that the operating frequency is fixed by the properties of the molecule and that the line width of the line used is so small, makes the device unsuitable for general application as a microwave amplifier. It has found many uses as a stable oscillator and the remark contained in this paper about the frequency stability has been borne out in practice.

Further details were given in *Phys. Rev.* **99**, 1253 and 1264 (1955).

PAPER 3*

Molecular Microwave Oscillator and New Hyperfine Structure in the Microwave Spectrum of NH³†

J. P. GORDON, H. J. ZEIGER‡ and C. H. TOWNES

Department of Physics, Columbia University, New York, New York

(Received May 5, 1954)

AN EXPERIMENTAL device, which can be used as a very high resolution microwave spectrometer, a microwave amplifier, or a very stable oscillator, has been built and operated. The device, as used on the ammonia inversion spectrum, depends on the emission of energy inside a high-Q cavity by a beam of ammonia molecules. Lines whose total width at half-maximum is six to eight kilocycles have been observed with the device operated as a spectrometer. As an oscillator, the apparatus promises to be a rather simple source of a very stable frequency.

A block diagram of the apparatus is shown in Fig. 1. A beam of ammonia molecules emerges from the source and enters a system of focusing electrodes. These electrodes establish a quadrupolar cylindrical electrostatic field whose axis is in the direction of the beam. Of the inversion levels, the upper states experience a radial inward (focusing) force, while the lower states see a radial outward force. The molecules arriving at the cavity are then virtu-

* *Phys. Rev.* **95**, 282–4 (1954).

† Work supported jointly by the Signal Corps, the U.S. Office of Naval Research, and the Air Force.

‡ Carbide and Carbon post-doctoral Fellow in Physics, now at Project Lincoln, Massachusetts Institute of Technology, Cambridge, Massachusetts.

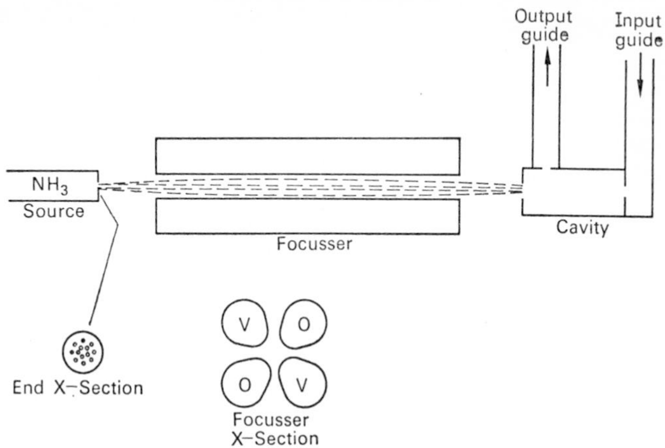

FIG. 1. Block diagram of the molecular beam spectrometer and oscillator.

ally all in the upper states. Transitions are induced in the cavity, resulting in a change in the cavity power level when the beam of molecules is present. Power of varying frequency is transmitted through the cavity, and an emission line is seen when the klystron frequency goes through the molecular transition frequency.

If the power emitted from the beam is enough to maintain the field strength in the cavity at a sufficiently high level to induce transitions in the following beam, then self-sustained oscillations will result. Such oscillations have been produced. Although the power level has not yet been directly measured, it is estimated at about 10^{-8} watt. The frequency stability of the oscillation promises to compare favorably with that of other possible varieties of "atomic clocks".

Under conditions such that oscillations are not maintained, the device acts like an amplifier of microwave power near a molecular resonance. Such an amplifier may have a noise figure very near unity.

High resolution is obtained with the apparatus by utilizing the directivity of the molecules in the beam. A cylindrical copper cavity was used, operating in the $TE011$ mode. The molecules, which travel parallel to the axis of the cylinder, then see a field which varies in amplitude as $\sin(\pi x/L)$, where x varies from 0 to L. In particular, a molecule traveling with a velocity v sees a field varying with time as $\sin(\pi vt/L)\sin(\Omega t)$, where Ω is the frequency of the rf field in the cavity. A Fourier analysis of this field, which the molecule sees from $t = 0$ to $t = L/v$, gives a frequency distribution whose amplitude drops to 0.707 of its maximum at points separated by a $\Delta\nu$ of $1.2v/L$. The cavity used was twelve centimeters long, and the most probable velocity of ammonia molecules in a beam at room temperature is 4×10^4 cm/sec. Since the transition probability is proportional to the square of the field amplitude, the resulting line should have a total width at half-maximum given by the above expression, which in the present case is 4 kc/sec. The observed line width of 6–8 kc/sec is close to this value.

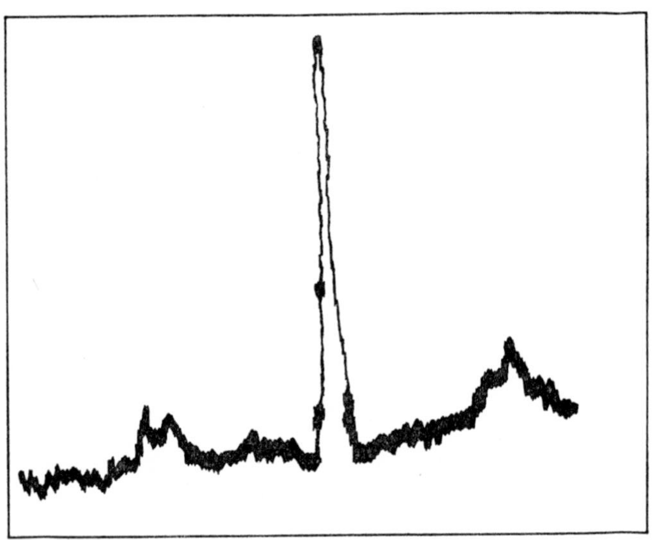

FIG. 2. A typical oscilloscope photograph of the NH_3, $J = K = 3$ inversion line at 23 870 Mc/sec, showing the resolved magnetic satellites. Frequency increases to the left.

The hyperfine structure of the ammonia inversion transitions for $J = K = 2$ and $J = K = 3$ has been examined, and previously unresolved structure due to the reorientation of the hydrogen spins has been observed. Figure 2 is a typical scope photograph of these new magnetic satellites on the 3,3 line. The observed spectra for the 3,3 line is shown in Fig. 3, which contains all the observed

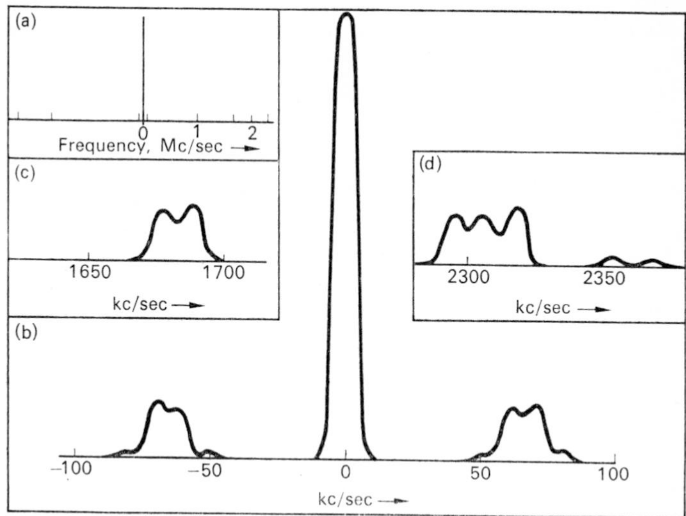

FIG. 3. The observed hyperfine spectrum of the 3,3 inversion line. (*a*) Complete spectrum, showing the spacings of the quadrupole satellites. (*b*) Main line with magnetic satellites. (*c*) Structure of the inner quadrupole satellites. (*d*) Structure of the outer quadrupole satellites. The quadrupole satellites on the low-frequency side of the main line are the mirror images of those shown, which are the ones on the high-frequency side.

hyperfine structure components, including the quadrupole reorientation transitions of the nitrogen nucleus, which have been previously observed as single lines.

Within the resolution of the apparatus, the hyperfine structures of the upper and lower inversion levels are identical, as evidenced by the fact that the main line is not split. Symmetry considerations require that the hydrogen spins be in a symmetric state under 120-degree rotations about the molecular axis. Thus for the 3,3 state, $I_H = 3/2$, and one expects each of the quadrupole levels to be further split into four components by the interaction of the hydrogen magnetic moments with the various magnetic fields of the molecule. At the present writing, the finer details of the expected magnetic splittings have not been worked out.

This type of apparatus has considerable potentialities as a more general spectrometer. Since the effective dipole moments of molecules depend on their rotational state, some selection of rotational states could be effected by such a focuser. Similarly, a focuser using magnetic fields would allow spectroscopy of atoms. Sizable dipole moments are required for a strong focusing action, but within this limitation, the device may prove to have a fairly general applicability for the detection of transitions in the microwave region.

The authors would like to acknowledge the expert help of Mr. T. C. Wang during the latter stages of this experiment.

Comments on Paper 4

THE authors introduce another method of obtaining the inversion necessary to make an amplifier or oscillator by showing that the effect of a strong pumping field in a three-level system can produce an inversion.

It is interesting to note that they point out that the inversion will be produced between the upper or lower levels depending on the positions of the three levels and will be greatest when the difference between the pump and signal frequencies is large as has been shown in Chapter 1. They suggest a number of systems which might be suitable, based on the various rotational and vibrational levels of molecules. Though these systems would make amplifiers, they would lack the tunability of the paramagnetic system considered by Bloembergen in the next paper.

These molecular systems have never been used in practice.

PAPER 4*

Possible Methods of Obtaining Active Molecules for a Molecular Oscillator

N. G. BASOV and A. M. PROKHOROV

P. N. Lebedev Institute of Physics, Academy of Sciences, U.S.S.R.

(Submitted to *J.E.T.P.* editor November 1, 1954)

J. Exper. Theoret. Phys. U.S.S.R. **28,** 249–250 (February, 1955)

As WAS shown in reference 1, one must use molecular beams in order to make a spectroscope with high resolving power. In this reference the possibility of constructing a molecular oscillator was investigated. Active molecules needed for self-excitation in the molecular oscillator were to be obtained by deflecting the molecular beam through inhomogeneous electric or magnetic fields. This method of obtaining active molecules has already been employed in the construction of a molecular oscillator.[2]

There is yet another way of obtaining active molecules, namely, pre-exposure of the molecular beam to auxiliary high frequency fields which induce resonance transitions between different levels of the molecules. In Fig. 1 and Fig. 2 we illustrate the possible variants which utilize an exciting irradiation of frequency ν_{ex} for populating the upper level in order to obtain a scheme of self-excitation with the frequency ν_g.

In one case, illustrated by Fig. 1, active molecules in level 1 are obtained at the expense of molecules in level 3 through transitions induced by the high frequency field. If the high frequency field

* *J. Exp. Theoret. Phys.* (*U.S.S.R.*) **1,** 184 (1954).

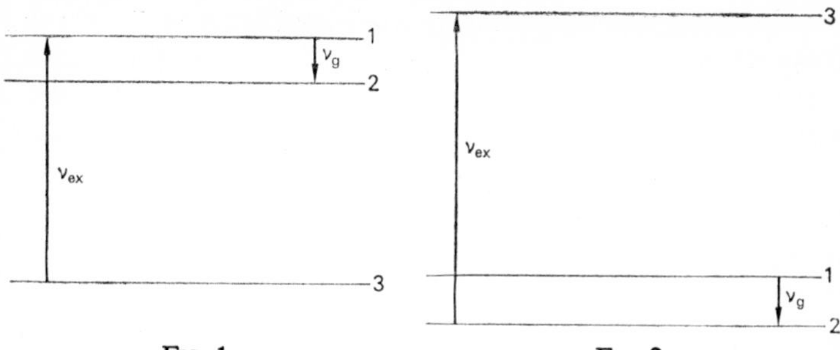

FIG. 1. FIG. 2.

possesses sufficient energy, so that the effect nears saturation, then the number of active molecules equals

$$\tfrac{1}{2}(N_3 - N_1) + N_1 - N_2, \tag{1}$$

where N_i is the number of the ith level.

The number of active molecules in level 1 increases with an increase of the energy difference between the first and third level relative to the energy difference between the first and the second levels. One must consider that the number of molecules in the levels is determined, in the case of thermodynamic equilibrium, by the Boltzmann factor.

$$N_i \sim e^{-E_i/kT}, \tag{2}$$

where E_i is the energy of the ith level and T is the absolute temperature of the molecular beam.

These considerations are valid for the case illustrated in Fig. 2. Here, however, instead of an increase of the number of molecules in level 1, we have a decrease of the number in level 2. The number of active molecules equals, in this case,

$$\tfrac{1}{2}(N_2 - N_3) + N_1 - N_2. \tag{3}$$

The method presented herein can be used, for example, in the following cases.

1) Levels 1 and 2 appear as neighboring rotational levels belonging to one and the same vibrational state of the molecule, with level 3 belonging to a neighboring vibrational state. In this case

the rotational quantum number of this level (level 3) differs from that of levels 1 and 2 by $\Delta J = 0, \pm 1$.

It is convenient to use the transitions between the vibrational levels for which $\Delta J = \pm 1$, since this case does not impose too strict a requirement for the exciting irradiation to be monochromatic. Since transitions between vibrational levels fall in the infra-red region of the spectrum for most molecules, the exciting irradiation must belong to this frequency range. However, infra-red thermal sources in existence at the present time have insufficient power to produce a saturation effect.

2) Levels 1, 2, and 3 are rotational levels of the molecule with asymmetric rotational momentum.

3) Levels 1 and 2 are hyperfine structures belonging to a given rotational state, and level 3 is a hyperfine level of a neighboring rotational level.

4) Levels 1 and 2 are specified by an inversion doublet belonging to a rotational level, and level 3 is one of the inversion levels of a neighboring rotational state.

The method presented here can be used to obtain a sufficient number of active molecules for the purpose of constructing a low frequency molecular oscillator.

Translated by A. Skumanich.

References

(1) N. G. Basov and A. M. Prokhorov, *J. Exper. Theoret. Phys. U.S.S.R.* **27**, 282 (1954).

(2) C. N. Townes *et al., Phys. Rev.* **95**, 282 (1954).

PAPER 5*

Proposal for a New Type Solid State Maser†

N. BLOEMBERGEN

Cruft Laboratory, Harvard University, Cambridge, Massachusetts

(Received July 6, 1956)

Summary

The Overhauser effect may be used in the spin multiplet of certain paramagnetic ions to obtain a negative absorption or stimulated emission at microwave frequencies. The use of nickel fluosilicate or gadolinium ethyl sulfate at liquid helium temperature is suggested to obtain a low noise microwave amplifier or frequency converter. The operation of a solid state maser based on this principle is discussed.

Townes and co-workers[1, 2] have shown that microwave amplification can be obtained by stimulated emission of radiation from systems in which a higher energy level is more densely populated than a lower one. In paramagnetic systems an inversion of the population of the spin levels may be obtained in a variety of ways. The "180° pulse" and the "adiabatic rapid passage" have been extensively applied in nuclear magnetic resonance. Combrisson and Honig[2] applied the fast passage technique to the two electron spin levels of a *P* donor in silicon, and obtained a noticeable power amplification.

Attention is called to the usefulness of power saturation of one transition in a multiple energy level system to obtain a change of

* *Phys. Rev.* **104,** 324–7 (1956).

† Supported by the Joint Services.

(1) Gordon, Zeiger and Townes, *Phys. Rev.* **99,** 1264 (1955).

(2) Combrisson, Honig and Townes, *Compt. rend.* **242,** 2451 (1956).

sign of the population difference between another pair of levels. A variation in level populations obtained in this manner has been demonstrated by Pound.[3] Such effects have since acquired wide recognition through the work of Overhauser.[4]

Consider for example a system with three unequally spaced energy levels, $E_3 > E_2 > E_1$. Introduce the notation,

$$h\nu_{31} = E_3 - E_1 \quad h\nu_{32} = E_3 - E_2 \quad h\nu_{21} = E_2 - E_1.$$

Denote the transition probabilities between these spin levels under the influence of the thermal motion of the heat reservoir (lattice) by

$$w_{12} = w_{21} \exp(-h\nu_{21}/kT), \quad w_{13} = w_{31} \exp(-h\nu_{31}/kT),$$
$$w_{23} = w_{32} \exp(-h\nu_{32}/kT).$$

The w's correspond to the inverse of spin lattice relaxation times. Denote the transition probability caused by a large saturating field $H(\nu_{31})$ of frequency ν_{31} by W_{13}. Let a relatively small signal of frequency ν_{32} cause transitions between levels two and three at a rate W_{32}. The numbers of spins occupying the three levels n_1, n_2, and n_3, satisfy the conservation law

$$n_1 + n_2 + n_3 = N.$$

For $h\nu_{32}/kT \ll 1$ the populations obey the equations:[5]

$$\frac{dn_3}{dt} = w_{13}\left(n_1 - n_3 - \frac{N}{3}\frac{h\nu_{31}}{kT}\right) + w_{23}\left(n_2 - n_3 - \frac{N}{3}\frac{h\nu_{32}}{kT}\right)$$
$$+ W_{31}(n_1 - n_3) + W_{32}(n_2 - n_3),$$

$$\frac{dn_2}{dt} = w_{23}\left(n_3 - n_2 + \frac{N}{3}\frac{h\nu_{32}}{kT}\right) + w_{21}\left(n_1 - n_2 - \frac{N}{3}\frac{h\nu_{21}}{kT}\right)$$
$$+ W_{32}(n_3 - n_2),$$

(3) R. V. Pound, *Phys. Rev.* **79**, 685 (1950).

(4) A. W. Overhauser, *Phys. Rev.* **92**, 411 (1953).

(5) In case $h\nu_{13} \sim kT$, the Boltzmann exponential factors cannot be approximated by the linear terms. The algebra becomes more involved without changing the character of the effect.

$$\frac{dn_1}{dt} = w_{13}\left(n_3 - n_1 + \frac{N}{3}\frac{h\nu_{31}}{kT}\right) + w_{21}\left(n_2 - n_1 + \frac{N}{3}\frac{h\nu_{21}}{kT}\right) - W_{31}(n_1 - n_3). \tag{1}$$

In the steady state the left-hand sides are zero. If the saturating field at frequency ν_{31} is very large, $W_{31} \gg W_{32}$ and w's, the solution is obtained

$$n_1 - n_2 = n_3 - n_2 = \frac{1}{3}\frac{hN}{kT}\frac{-w_{23}\nu_{32} + w_{21}\nu_{21}}{w_{23} + w_{12} + W_{32}}. \tag{2}$$

This population difference will be positive, corresponding to negative absorption or stimulated emission at the frequency ν_{32}, if

$$w_{21}\nu_{21} > w_{32}\nu_{32}. \tag{3}$$

If the opposite is true, stimulated emission will occur at the frequency ν_{21}. The following discussion could easily be adapted to this situation. The power emitted by the magnetic specimen is

$$P_{\text{magn}} = \frac{Nh^2\nu_{32}}{3kT}\frac{(w_{21}\nu_{21} - w_{32}\nu_{32})W_{32}}{w_{23} + w_{12} + W_{32}}. \tag{4}$$

For a magnetic resonance line with a normalized response curve $g(\nu)$ and $g(\nu_{\max}) = T_2$, the transition probability at resonance is given by

$$W_{32} = \hbar^{-2} \mid (2 \mid M_x \mid 3) \mid^2 H_s^{\,2}(\nu_{32}) T_2. \tag{5}$$

For simplicity it has been assumed that the signal field $H(\nu_{32})$ is uniform in the x direction over the volume of the sample. A similar expression holds for W_{31}.

For the moment we shall restrict ourselves to the important case that the signal excitation at frequency ν_{32} is small, $W_{32} \ll w_{23} + w_{31}$. No saturation effects at this transition occur and a magnetic quality factor can be defined by

$$-1/Q_{\text{magn}} = \frac{4P_{\text{magn}}}{\nu_{23}\langle H^2(\nu_{32})\rangle_{\text{Av}} V_c}. \tag{6}$$

Q_{magn} is negative for stimulated emission, $P_{\text{magn}} > 0$. V_c is the volume of the cavity, and $\langle H^2 \rangle_{\text{Av}}$ represents a volume average over the cavity. The losses in the cavity, exclusive of the magnetic losses or gains in the sample, are described by the unloaded quality factor Q_0. The external losses from the coupling to a wave guide or coaxial line are described by Q_e. Introduce the voltage standing wave ratio β for the cavity tuned to resonance,

$$\beta = (Q_e/Q_0) + (Q_e/Q_{\text{magn}}).$$

The ratio of reflected to incident power is

$$\frac{P_r}{P_i} = \frac{(1-\beta)^2}{(1+\beta)^2}.$$

There is a power gain or amplification, when β is negative or, $-Q_{\text{magn}}{}^{-1} > Q_0{}^{-1}$. Oscillation will occur when

$$-Q_{\text{magn}}{}^{-1} > Q_0{}^{-1} + Q_e{}^{-1} = Q_L{}^{-1},$$

where Q_L is the "loaded Q". The amplitude of the oscillation will be limited by the saturation effect, embodied by the W_{32} in the denominator of Eq. (4). The absolute value of $1/Q_{\text{magn}}$ decreases as the power level increases. In the oscillating region the device will act as a microwave frequency converter. Power input is at the frequency ν_{13}, a smaller power output at the frequency ν_{23}. The balance of power is dissipated in the form of heat through the spin-lattice relaxation and through conduction losses in the cavity walls. For $-Q_{\text{magn}} = Q_L$, $\beta = -1$, and the amplification factor would be infinite. The device will act as a stable c.w. amplifier at frequency ν_{23}, if

$$Q_0{}^{-1} + Q_e{}^{-1} > -Q_{\text{magn}}{}^{-1} > Q_0{}^{-1}. \tag{7}$$

The choice of paramagnetic substance is largely dependent on the existence of suitable energy levels and the existence of matrix elements of the magnetic moment operator between the various spin levels. The absorption and stimulated emission process depend directly on this operator, but the relaxation terms (w) also depend on the spin angular momentum operator via spin–orbit coupling terms. It is essential that all off-diagonal elements between the three spin levels under consideration be nonvanishing. This can be achieved by putting a paramagnetic salt with a crystalline field splitting δ in a magnetic field, which makes an angle with the crystalline field axis. The magnitude of the field is such that the Zeeman energy is comparable to the crystalline field splitting. In this case the states with magnetic quantum numbers m_s are all scrambled. This situation is usually avoided to unravel paramagnetic resonance spectra, but occasionally "forbidden lines" have been observed, indicating mixing of the m_s states. For our purposes the mixing up of the spin states by Zeeman and crystalline field interactions of comparable magnitude is essential. The energy levels and matrix elements of the spin angular momentum operator can be obtained by a numerical solution of the determinantal problem of the spin Hamiltonian.[(6)] The number of electron spin levels may be larger than three. One may choose the three levels between which the operation will take place. The analysis will be similar, but algebraically more complicated. One has a considerable amount of freedom by the choice of the external dc magnetic field, to adjust the frequencies ν_{23} and ν_{13} and to vary the values of the inverse relaxation times w. It is advisable—although perhaps not absolutely necessary—to operate at liquid helium temperature. This will give relatively long relaxation times (between 10^{-2} and 10^{-4} sec), and thus keep the power requirements for saturation down. The factor T in the denominator of Eq. (4) will also increase the emission at low temperature. Although the order of magnitude of the w's is known through the work of Leiden school,[(7)] there is only one instance where w's have been measured for some individual transitions.[(8)] Van Vleck's[(9)] theory of paramagnetic

[(6)] See, e.g. B. Bleaney and K. H. W. Stevens, *Repts. Progr. in Phys.* **16**, 108 (1953).

[(7)] See, e.g. C. J. Gorter, *Paramagnetic Relaxation* (Elsevier Publishing Company, Amsterdam, 1948).

[(8)] A. H. Eschenfelder and R. T. Weidner, *Phys. Rev.* **92**, 869 (1953).

[(9)] J. H. Van Vleck, *Phys. Rev.* **57**, 426 (1940).

relaxation should be extended to the geometries envisioned in this paper. If a Debye spectrum of the lattice vibrations is assumed, the relaxation times will increase with decreasing frequency at liquid helium temperature, where Raman processes are negligible. This implies that the condition (3) should be easily realizable when $\nu_{32} < \nu_{21}$.

Important applications as a microwave amplifier could, e.g., be obtained for $\nu_{32} = 1420$ Mc/sec, corresponding to the interstellar hydrogen line, or to another relatively low microwave frequency used in radar systems. The frequency ν_{31} could be chosen in the X band, $\nu_{31} = 10^{10}$ cps. To obtain well scrambled states with these frequency splittings one should have crystalline field splittings between 0.03 cm^{-1} and 0.3 cm^{-1}. Paramagnetic crystals which are suggested by these considerations are nickel fluosilicate[(10)] and gadolinium ethyl sulfate.[(11)] These crystals have the additional advantage that all magnetic ions have the same crystalline field and nuclear hyperfine splitting is absent, thus keeping the total number of possible transitions down. The use of magnetically dilute salts is indicated to reduce the line width, increase the value of T_2 in Eq. (5) and to separate the individual resonance transitions.

A single crystal 5% Ni 95% Zn Si $F_6 \cdot 6H_2O$ has a line width of 50 oersted ($T_2 = 1.2 \times 10^{-9}$ sec) and an average crystalline field splitting $\delta = 0.12$ cm^{-1} for the Ni^{++} ions. With an effective spin value $S = 1$ there are indeed three energy levels of importance. The spin lattice relaxation time is about 10^{-4} sec at 2°K as measured in a saturation experiment by Meyer.[(12)] Further dilution does not decrease the line width, as there is a distribution of crystalline fields in the diluted salt.

A single crystal of 1% Gd 99% La $(C_2H_5SO_4)_3 \cdot 9H_2O$ has an effective spin $S = 7/2$. In zero field there are four doublets

(10) R. P. Penrose and K. H. W. Stevens, *Proc. Phys. Soc.* (*London*) **A63**, 29 (1949).

(11) Bleaney, Scovil and Trenam, *Proc. Roy. Soc.* (*London*) **A223**, 15 (1954).

(12) J. W. Meyer, Lincoln Laboratory Report 1955 (unpublished).

separated respectively by $\delta = 0.113\ \text{cm}^{-1}$, $0.083\ \text{cm}^{-1}$, and $0.046\ \text{cm}^{-1}$ as measured at 20°K. These splittings are practically independent of temperature. The line width is 7 oersteds due to the distribution of local fields arising from the proton magnetic moments. This width could be reduced by a factor three by using the deuterated salt. The relaxation time is not known, but should be about the same as in other Gd salts,[7] which give $T_1 \sim 10^{-2}$ sec at 2°K.

In the absence of detailed calculations for the relaxation mechanism, we shall take $w_{12} = w_{13} = w_{32} = 10^4\ \text{sec}^{-1}$ for the nickel salt and equal to $10^2\ \text{sec}^{-1}$ for the gadolinium salt. The matrix elements $(2 \mid M_x \mid 3)$, etc., can be calculated exactly by solving the spin determinant. For the purpose of judging the operation of the maser using these salts, we shall take the off-diagonal elements of magnetic moment operator simply equal to $g\beta_0$, where $g = 2$ is the Landé spin factor and β_0 is the Bohr magneton. For the higher spin value of the Gd^{+++} some elements will be larger but this effect is offset by the distribution of the ions over eight rather than three spin levels. Take $T = 2$°K and $Q_0 = 10^4$, which is readily obtained in a cavity of pure metal at this temperature. A coaxial cavity may be used which has a fundamental mode resonating at the frequency $\nu_{32} = 1.42 \times 10^9$ cps and a higher mode resonating at $\nu_{31} \approx 10^{10}$ cps. Take the volume of the cavity $V = 60\ \text{cm}^3$ and $H_s^{\,2} = 6\langle H^2 \rangle_{\text{Av}}$. If these values are substituted in Eqs. (4)–(6), the condition (7) for amplification is satisfied if $N > 3 \times 10^{18}$ for nickel fluosilicate and $N > 3 \times 10^{17}$ for gadolinium ethyl sulfate ($N > 10^{17}$ for the deuterated salt). The minimum required number of Ni^{++} ions are contained in $0.02\ \text{cm}^3$ of the diluted nickel salt. The gadolinium salt, diluted to 1% Gd, contains the required number in about the same volume. The critical volume is only $0.006\ \text{cm}^3$ for the deuterated salt. Crystals of appreciably larger size can still be fitted conveniently in the cavity. A c.w. amplifier or frequency converter should therefore be realizable with these substances. A larger amount of power can be handled by these crystals than by the *P* impurities in silicon which have a very long relaxation time, and require an intermittent operation, and where it is harder to get the required number of spins in the cavity.

So far we have assumed that the width corresponds to the inverse of a true transverse relaxation time T_2. Actually the width $1/T_2^*$ is due to an internal inhomogeneity broadening with normalized distribution $h(\nu)$ and $h(\nu_{max}) \approx T_2^*$ in both cases. The response curve for a single magnetic ion is probably very narrow indeed, $g(\nu_{max}) = T_2 \approx T_1$, and $T_1 = 10^{-4}$ should be used in Eq. (5) rather than $T_2^* = 1.2 \times 10^{-9}$ sec. The response to a weak threshold signal at ν_{32} now originates, however, from a small fraction of the magnetic ions. If $\gamma H(\nu_{32}) < 1/T_1 \approx 10^4$ cps, then only T_2^*/T_1 of the ions contribute to the stimulated emission and the net result is the same as calculated above. In most applications the incoming signal will be so weak that this situation will apply, even with a power amplification of 30 or 40 db.

For use as an oscillator or high level amplifier with a field $H(\nu_{32})$ in the cavity larger than $1/\gamma T_1$, one has essentially complete saturation ($W_{32} \gg w_{23}+w_{13}$) in Eq. (4) for those magnetic ions lying in a width $2\pi\Delta\nu = \gamma H(\nu_{32})$ in the distribution $h(\nu)$. One has then for the power emitted instead of Eqs. (4) and (5)

$$P_{\text{magn}} = \frac{h^2\nu_{32}}{3kT} N(-w_{32}\nu_{32}+w_{21}\nu_{21})\gamma H(\nu_{32})T_2^*. \tag{8}$$

The power is proportional to the amplitude of the radio frequency field rather than its square. This effect has been discussed in more detail by Portis.[13] It will limit the oscillation or amplification to an amplitude which can be calculated by using Eq. (8) in conjunction with Eqs. (6) and (7).

The driving field $H(\nu_{31})$ will necessarily have to satisfy the condition $\gamma H(\nu_{31}) > w_{31} = T_1^{-1}$ to obtain saturation between levels 1 and 3. The power absorbed in the crystal will be proportional to the amplitude $H(\nu_{31})$, and is in order of magnitude given by

$$P_{\text{abs}} \sim N\frac{h^2\nu_{13}^{\ 2}}{3kT} w_{13}\gamma H(\nu_{31})T_2^*. \tag{9}$$

[13] A. M. Portis, *Phys. Rev.* **91**, 1071 (1953).

This equation loses its validity if $\gamma H(\nu_{31}) > T_2^{*-1}$. In this case the whole line would be saturated, but such excessive power levels will not be used. For $T_1^{-1} < \gamma H(\nu_{31}) < T_2^{*-1}$, the effective band width of the amplifier is determined by $H(\nu_{31})$. It is about 0.5 Mc/sec for $H(\nu_{31}) = 0.2$ oersted. The power dissipated in a specimen of fluosilicate ten times the critical size is 0.5 milliwatt under these circumstances. For the gadolinium salt, also ten times the critical size, either deuterated or not, the dissipation is only 0.005 milliwatt. There should be no difficulty in carrying this amount out of the paramagnetic crystal without excessive heating. The power dissipation in the walls under these conditions will be 5 milliwatts. Liquid helium will boil off at the rate of only 0.01 cc/min due to heating in the cavity. Since helium is superfluid at 2°K, troublesome vapor bubbles in the cavity are eliminated.

The noise power generated in this type of amplifier should be very low. The cavity with the paramagnetic salt can be represented by two resonant coupled circuits as discussed by Bloembergen and Pound.[(14)] Noise generators are associated with the losses in the cavity walls, kept at 2°K, and with the paramagnetic spin absorption which is described by an effective spin temperature, associated with the distribution of the spin population. The absolute value of this effective temperature also has the order of magnitude of 1°K. The input is from an antenna, which sees essentially the radiation temperature of interstellar space. Reflected power is channeled by a circulating nonreciprocal element[(15)] into a heterodyne receiver, or, if necessary, into a second stage Maser cavity. The circulator makes the connection: antenna → maser cavity → heterodyne receiver → dummy load → antenna. If the antenna is not well matched, the dummy load may be a matched termination kept at liquid helium temperature to prevent extra power from entering the cavity. The input arm of the cavity at frequency ν_{31} will be beyond cutoff for the frequency ν_{32}. The coaxial line passing the signal at ν_{32} between cavity and circulator

(14) N. Bloembergen and R. V. Pound, *Phys. Rev.* **95,** 8 (1954).

(15) C. L. Hogan, *Bell System Tech. J.* **31,** 1 (1952).

will contain a rejection filter at frequency ν_{31} to prevent overloading and noise mixing at the mixer crystal of the super heterodyne receiver.

It may be concluded that the realization of a low-noise c.w. microwave amplifier by saturation of a spin level system at a higher frequency seems promising. The device should be particularly suited for detection of weak signals at relatively long wavelengths, e.g., the 21-cm interstellar hydrogen radiation. It may also be operated as a microwave frequency converter, capable of handling milliwatt power. More detailed calculations and design of the cavity are in progress.

PAPER 6*

Paramagnetic Fine Structure Spectrum of Cr^{+++} in a Single Ruby Crystal†

J. E. Geusic

Department of Physics, The Ohio State University, Columbus, Ohio
(Received February 20, 1956)

Summary

The fine structure spectrum of the Cr^{+++} paramagnetic resonance line in a single ruby crystal has been studied at 300°K. From the measurements it is found that the zero-field splitting of the spin levels is $2(0.193 \pm 0.001)$ cm^{-1} and the spectroscopic splitting factors are $g_{\parallel} = 2.003 \pm 0.006$ and $g_{\perp} = 2.00 \pm 0.02$.

I. Introduction

Magnetic resonance absorption at 9309 Mc/sec has been observed in a single ruby crystal at 300°K as a function of the static magnetic field for several orientations of the crystalline *c*-axis relative to the magnetic field. The resonance absorption has been interpreted as due to Cr^{+++} ions in a crystalline electric field of trigonal symmetry. From the measurements the splitting of the ground state has been inferred along with $g_{\parallel}$ and $g_{\perp}$.

Resonance absorption due to transitions between the split ground-state levels of the Cr^{+++} ion was also observed in approximately zero field at 11 593 Mc/sec.

* *Phys. Rev.* **102,** 1252–3 (1956).

† This work was supported in part by a contract between the Office of Scientific Research of the Air Research and Development Command, Baltimore, Maryland and The Ohio State University Research Foundation.

The basic structure of ruby is rhombohedral[1] in which all Cr and Al atoms are in equivalent sites whose symmetry is trigonal; consequently the internal electric fields also have trigonal symmetry about the crystalline *c*-axis. The effect of the crystalline field perturbation on the $^4F_{\frac{3}{2}}$ state of the free chromic ion is to split this state into two twofold orbitally degenerate states and three onefold orbital states all of which are $(2S+1)$-fold degenerate with respect to the spin. In the interpretation of the data it has been assumed that an orbital singlet is the lowest energy state and that the $\lambda(\mathbf{L}\cdot\mathbf{S})$ perturbation then removes partially the spin degeneracy giving two twofold degenerate spin levels in zero magnetic field. On the basis of these assumptions a spin Hamiltonian can be deduced which must show the same symmetry as the crystalline electric field. For the case of trigonal symmetry[2] it takes the form

$$H = DS_z^{\,2}+\beta[g_{\parallel}S_zH_z+g_{\perp}(S_xH_x+S_yH_y)], \tag{1}$$

provided that the effect of $\lambda(\mathbf{L}\cdot\mathbf{S})$ is taken into account only as far as second order and that all nuclear interactions are neglected. The *z*-axis is chosen in (1) to coincide with the trigonal axis of the electric field; therefore, (1) can be written in the convenient form

$$H = DS_z^{\,2}+\beta H[g_{\parallel}S_z\cos\theta+g_{\perp}\sin\theta\{\tfrac{1}{2}e^{i\varphi}(S_x-iS_y) \\ +\tfrac{1}{2}e^{-i\varphi}(S_x+iS_y)\}], \tag{2}$$

where θ is the polar angle between the *z*-axis of the crystal and the externally applied magnetic field and φ is the azimuthal angle about the *z*-axis. The parameters D, $g_{\parallel}$ and $g_{\perp}$ can be obtained from measurements for $\theta = 0°$ and $\theta = 90°$.

II. Experimental Method

The experimental arrangement used is shown in Fig. 1. The method consisted in detecting the change in Q value of a rec-

(1) R. W. Wyckoff, *The Structure of Crystals* (The Chemical Catalog Company, Inc., New York, 1931).

(2) B. Bleaney and K. W. H. Stevens, *Rept. Progr. Phys.* **16,** 108 (1951).

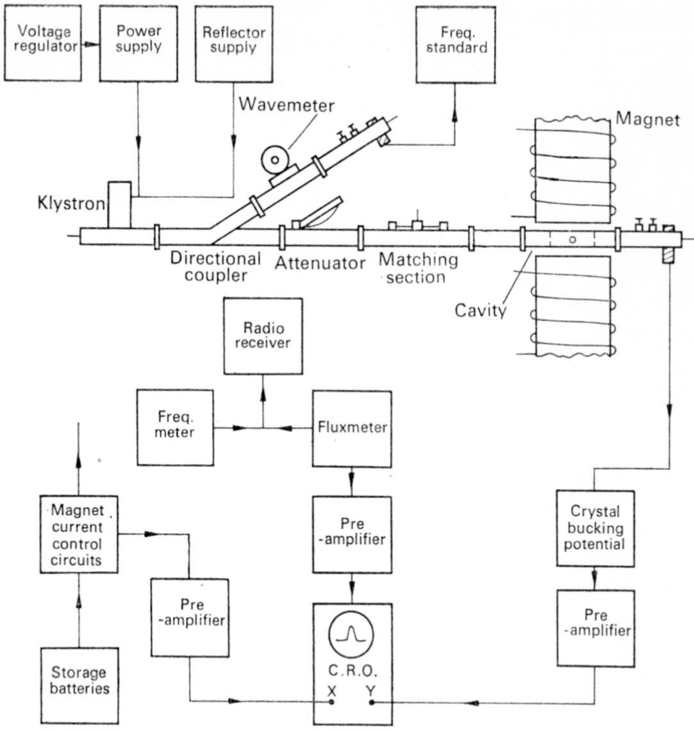

FIG. 1. Block diagram of the paramagnetic resonance spectrograph.

tangular cavity excited by a reflex klystron oscillator. The absorption of the microwave energy by the sample reduced the Q value of the cavity and this was detected as a change in the dc voltage across the crystal diode detector. This dc voltage was amplified and put onto the vertical plates of an oscilloscope. A small voltage proportional to the current through the magnet coils was applied to the horizontal plates of the oscilloscope, so that as the magnetic field was varied through a resonance a plot of absorbed power *versus* magnet current was made. A proton resonance flux meter was used to measure the magnetic field at the center of a line by superimposing the proton resonance on the paramagnetic resonance line. This was done by feeding the output of the fluxmeter into the time marker channel of the oscilloscope. During the measurements the klystron was monitored and measured by a secondary standard calibrated in terms of WWV.

The essentially zero-field measurements were made using an X-13 Varian klystron. In this situation the magnetic field was swept from $+100$ gauss to -100 gauss and the frequency shifted until the two oscilloscope traces of the low magnetic-field line, one corresponding to the positive value of the resonance field and the other corresponding to the negative value of the resonance field, were made to coincide.

The single crystal used in the experiments was held in the magnetic field by means of a device incorporating a divided circle directly readable to one minute of arc. The crystal was cemented to a polystyrene rod whose axis coincided with the axis of the divided circle and in such a manner that the c-axis was perpendicular to the axis of the graduated circle. The crystal and crystal mount were lined up with respect to the magnetic field by optical means so that the c-axis could be oriented parallel to and perpendicular to the field by rotations about the axis of the graduated circle.

III. Experimental Results

Two crystals were used in the measurements. One was a natural ruby crystal, which contained probably less than 1% of chromium, whose c-axis was determinable by visual observation. The other crystal was a synthetic ruby in the form of a slender cylindrical rod whose c-axis made an angle of 56° with the axis of the rod. The latter contained approximately 2% of chromium.

The position of the fine structure lines for both samples agreed rather well with one another; however, the measurements on the natural ruby were more reproducible because of its narrower lines.

The data for both samples could be best fitted with the Hamiltonian in (1) by setting

$$D = 0.193 \pm 0.001 \text{ cm}^{-1},$$
$$g_{\parallel} = 2.003 \pm 0.006,$$
$$g_{\perp} = 2.00 \pm 0.02.$$

This value of D agreed with the value obtained from the zero-field measurements. The agreement between experimental and calculated values of the resonance fields were within experimental error for $\theta = 0°$; however, for $\theta = 90°$ the agreement was only good to within 1 percent of the measured value. To check the agreement at intermediate angles. numerical tables compiled by Parker[(3)]

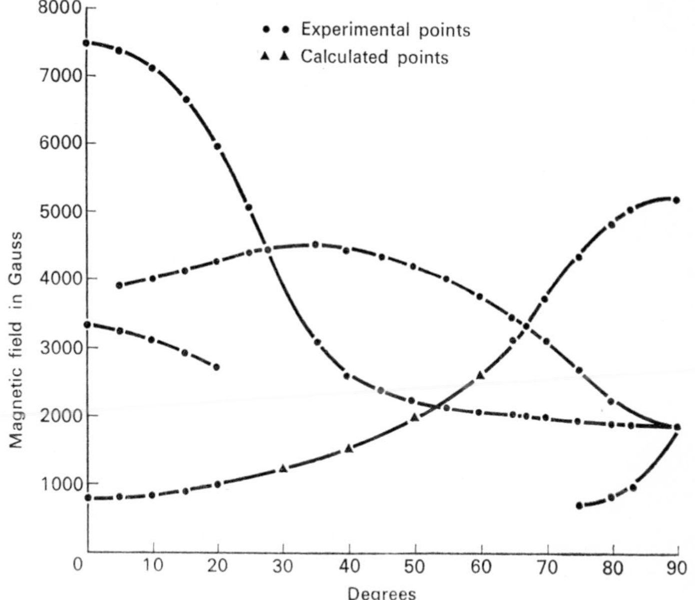

FIG. 2. A plot of the resonance fields at 9309 Mc/sec of the Cr^{+++} fine-structure spectrum as a function of the polar angle θ.

[(3)] P. Parker, *J. Chem. Phys.* (to be published).

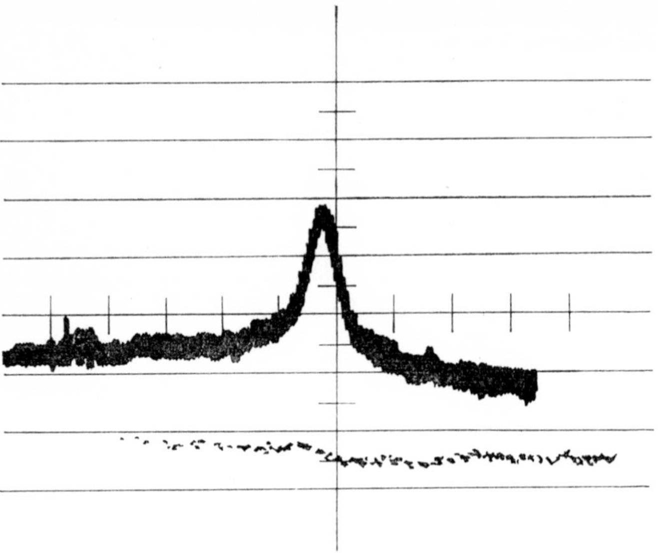

FIG. 3. A typical oscilloscope trace of one of the fine-structure lines observed in a single natural ruby crystal.

were used. For all angles the measured values agreed with calculated line positions to within 1 percent of the measured value. The variation of the fine-structure lines with polar angle θ is shown in Fig. 2. Some calculated points for the low-field line are shown at orientations where the intensity of the line was too small for observation. The line widths observed in the case of the natural ruby were roughly 45–70 gauss. A typical resonance line in the natural ruby is shown in Fig. 3.

IV. Conclusions

The resonance spectrum of the Cr^{+++} ion seems to be generally described by (1); however, some of the discrepancies between experimental and calculated values, which cannot be attributed to experimental error, may be due to the fact that fourth order terms of the spin operators have not been included in (1). It is felt that probably the addition of these higher order terms will not change appreciably the value of D obtained in these experiments but may shift the g values slightly.

V. Acknowledgments

The author wishes to express his appreciation to Professor Dudley Williams, Professor Jan Korringa, and Dr. L. Carlton Brown for their helpful discussions and to Professor Duncan McConnell and Linde Air Products Company for supplying the ruby crystals used in the experiment.

PAPER 7*

Paramagnetic Spectra of Substituted Sapphires—Part 1: Ruby†

E. O. SCHULZ-DU BOIS

(Manuscript received September 29, 1958)

Summary

The paramagnetic resonance properties of Cr^{+++} ions in Al_2O_3 (ruby) were investigated theoretically and experimentally in order to obtain information necessary for the application of this material as active material in a three-level solid-state maser (3LSSM). Numerically computed energy levels, together with their associated eigenvectors, are presented as a function of applied magnetic field for various orientations of the magnetic field with respect to the crystalline symmetry axis. A more detailed discussion is devoted to energy levels, eigenvectors and transition probabilities at angles 0°, 54.74° and 90°, where certain simple relations and symmetries hold. Paramagnetic spectra for signal frequencies between 5 and 24 kmc are shown; agreement between computed and measured resonance fields is satisfactory.

I. Introduction

Among the paramagnetic salts that have been used as active materials in three-level solid-state masers (3LSSM),[1, 2, 3] ruby shows rather desirable properties. While maser action of this material has been achieved at microwave signal frequencies of 3 to 10 kmc,[4] it should be possible to cover more than the whole centimeter microwave range. Perhaps even more important from a practical point of view are the bulk physical properties. Ex-

* *Bell Syst. Tech. J.* **38,** 271–90 (1959).

† This work is partially supported by the Signal Corps under Contract Number DA–36–039 sc–73224.

tremely good heat conductivity at low temperatures allows handling of relatively high microwave power dissipation. Industrial growth of large single crystals by the flame fusion technique and machinability with diamond tools make it possible to fabricate long sections of ruby to very close tolerances, a necessity in traveling-wave maser (TWM) development. Also, ruby can be bonded to metals, thus allowing a high degree of versatility in maser structural design. While the use of ruby in 3LSSM, in particular in nonreciprocal TWM, will be described in forthcoming papers by members of Bell Telephone Laboratories, this paper is intended to give some background on paramagnetic resonance behavior of ruby.

In general, the paramagnetic resonance properties of an ion in a crystal can be completely described by a spin Hamiltonian containing a relatively small number of constants. In the case of ruby, these include the spectroscopic splitting factors parallel and perpendicular to the crystalline axis, $g_{\parallel}$ and $g_{\perp}$, the total spin $S = 3/2$ and the sign and magnitude of $2D$, the zero field splitting. Nuclear interactions can be neglected since the most abundant isotope, Cr^{52}, is nonmagnetic ($I = 0$) whereas the magnetic isotope, Cr^{53}, ($I = 3/2$) has small abundance (9.5 per cent) and leads to negligible line broadening only. Taking this into account, one can even predict on the basis of the total spin and the crystalline symmetry surrounding the Cr^{+++} ion that no other terms can occur in the spin Hamiltonian.

However, in order to predict operating conditions of this or other materials in a 3LSSM, it is necessary to know the separation of energy levels for supplying the proper pump and signal frequencies, the order of magnitude of the associated transition probabilities and perhaps other circumstances, such as coincidence of transition frequencies, which, by spin–spin interaction, may lead to shortening of the associated relaxation times (self-doping condition). In this paper, this information is evaluated by the formalism of the spin Hamiltonian and, at least in part, compared with experiment. The data presented graphically are intended to form an "atlas" of the ruby paramagnetic resonance properties.

In the paper which follows, some general viewpoints are presented on modes in which paramagnetic materials can be operated as active materials in a 3LSSM. In further papers, paramagnetic spectra of other substitutional ions such as Co^{++} and Fe^{+++} in sapphire will be presented in order to furnish sufficient information to find coincidences of transition frequencies of Cr^{+++} with Co^{++} or Fe^{+++} lines resulting in reduced relaxation times (impurity-doping condition).

For a derivation of the method of spin Hamiltonians, reference should be made to such review articles as those by Bleaney and Stevens[5] and Bowers and Owen.[6] Knowledge of the associated formalism is perhaps desirable but not necessary for utilization of the results reported in this paper. Briefly, the spin Hamiltonian describes the energy of a paramagnetic ion arising from interaction with host crystal environment and applied magnetic field. Obeying quantum laws, the ion can exist in one of several states associated with discrete energy levels. Transitions between such states can occur if the energy balance ΔE is supplied to or extracted from the ion. Given some probability for radiative transitions, these can be induced by applying a magnetic field of radio frequency $\nu = \Delta E/h$ (h = Planck's constant). If there are more transitions to the higher state, net absorption will be observed such as is normally observed with a spectrometer. If there are more transitions to the lower state, stimulated emission of energy will be observed such as is utilized for amplification in a 3LSSM.

II. The Spin Hamiltonian

The spin Hamiltonian of Cr^{+++} in Al_2O_3 was first published by Manenkov and Prokhorov,[7] and later by Geusic[8] and Zaripov and Shamonin.[9] It was given in the form

$$\mathscr{H} = g_{\parallel} H_z S_z + g_{\perp}(H_x S_x + H_y S_y) + D[S_z^2 - \tfrac{1}{3}S(S+1)]. \quad (1)$$

The effective spin $S = 3/2$ is identical with the true spin. All Cr^{+++} ions in the crystal lattice show identical paramagnetic behavior, with the magnetic z-axis being the same as the trigonal symmetry axis of the crystal. The best values for the constants seem to be

$$2D = -2D' = -0.3831 \pm 0.0002 \text{ cm}^{-1} = -11.493 \pm 0.006 \text{ kmc},$$
$$g_{\parallel} = 1.9840 \pm 0.0006,$$
$$g_{\perp} = 1.9867 \pm 0.0006.$$

While it is customary in spectroscopy to express energy in units of cm^{-1} omitting a factor hc (h = Planck's constant, c = velocity of light), units of kmc are used simultaneously, omitting a factor of $10^9\, h$, because this allows direct interpretation in observed spectra.

In particular, the negative sign of D was obtained by Geusic.(8) He deduced this from the fact that $g_{\parallel} < g_{\perp}$, since in less than half-filled d-shell ions, such as Cr^{+++}, the spin–orbit coupling term λ is positive, and D is given by $2D = \lambda(g_{\parallel} - g_{\perp})$. Sign and magnitude D are in agreement with results of low-temperature static susceptibility measurements by Bruger.(10) In this work also, the negative sign of D was confirmed by comparing the relative intensities of two lines at liquid nitrogen and helium temperatures.

The spin Hamiltonian (1) can more conveniently be written in spherical coordinates:

$$\mathscr{H} = g_{\parallel} H \cos\theta S_z + \tfrac{1}{2} g_{\perp} H \sin\theta (e^{-i\varphi} S_+ + e^{i\varphi} S_-) - D'[S_z^2 - \tfrac{1}{3} S(S+1)]. \quad (2)$$

Here $S_{\pm} = S_x \pm i S_y$. In both representations (1) and (2) the crystalline axis was chosen to be the z-axis. While the choice of reference system is immaterial to obtaining eigenvalues (energy levels), this choice shows up in the associated eigenvectors. The eigenvectors have no direct physical interpretation; they must be evaluated in order to obtain transition probabilities. The transition probabilities most naturally obtained from eigenvectors of the Hamiltonian (2) are those which correspond to excitation by RF magnetic fields whose polarization is either linear and parallel to, or circular and perpendicular to, the *crystalline axis*.

In 3LSSM design, however, it seems more appropriate to analyze the performance in terms of RF magnetic fields whose polarization is either linear and parallel to, or circular and perpendicular to, the *applied field.* The corresponding eigenvectors and transition probabilities can, of course, be obtained from those belonging to the Hamiltonian (2) by a 4-by-4 transformation matrix. But it is more efficient to obtain them directly through a transformation of the original spin Hamiltonian (1) or (2) into a coordinate system with the z-axis parallel to the applied field. The result of this transformation is

$$\begin{aligned}\mathscr{H} = {} & (g_{\parallel} \cos^2 \theta + g_{\perp} \sin^2 \theta)\beta H S_z \\ & - D'(\cos^2 \theta - \tfrac{1}{2} \sin^2 \theta)\,[S_z{}^2 - \tfrac{1}{3}S(S+1)] \\ & - D'\tfrac{1}{2} \cos \theta \sin \theta[e^{-i\varphi}(S_z S_+ + S_+ S_z) + e^{i\varphi}(S_z S_- + S_- S_z)] \\ & - D'\tfrac{1}{4} \sin^2 \theta(e^{-2i\varphi} S_+{}^2 + e^{2i\varphi} S_-{}^2). \end{aligned} \tag{3}$$

III. Energy Levels and Eigenvectors

From the Hamiltonian $\mathscr{H}$ (3), its energy eigenvalues W are found numerically by solving the fourth-order secular equation

$$\| \langle n \mid \mathscr{H} - W \mid m \rangle \| = 0,$$
$$n, m = 3/2,\ 1/2,\ -1/2,\ -3/2. \tag{4}$$

The eigenvalues W are functions of H and θ, but not of φ since, because of the symmetry of the Hamiltonian, rotation about the z-axis does not change the physical situation. On the following plots (left-hand sections of Figs. 1 through 11) diagrams of energy levels W (in units of kmc) are shown as a function of applied field H (in units of kilogauss). Plots are given for angles θ from 0° to 90° in steps of 10° and, in addition, for 54.74°.

Also, by change of scales, dimensionless eigenvalues $y = W/D'$ are shown as functions of the dimensionless quantity $x = G/D'$, where

$$G = (g_{\parallel} \cos^2 \theta + g_{\perp} \sin^2 \theta)\beta H.$$

This dimensionless representation facilitates computations and reveals more clearly symmetries and singular relations in the energy level scheme. It also permits the use of the same diagrams for ions having the same Hamiltonian but different zero field

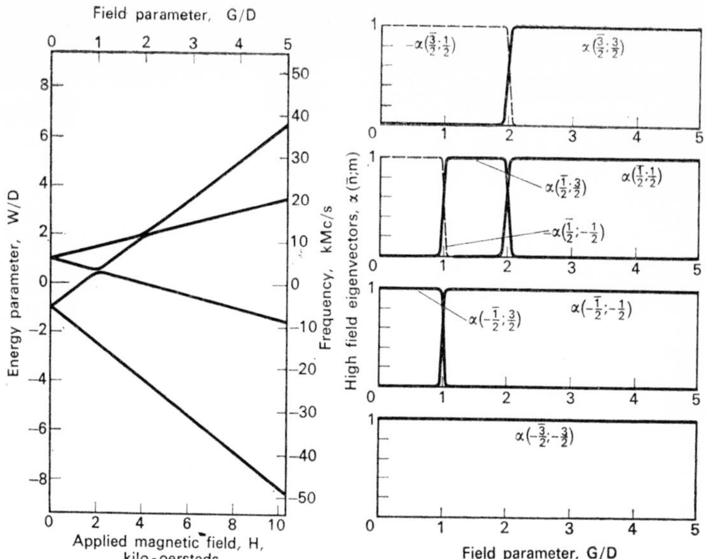

FIG. 1. Energy levels and eigenvectors of the Cr^{+++} paramagnetic ion in ruby at angle $\theta = 0°$ between crystalline symmetry axis and applied magnetic field.

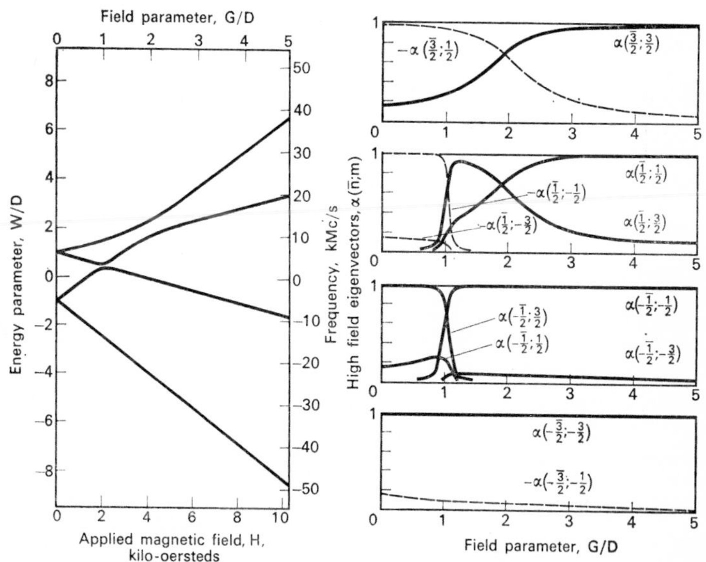

FIG. 2. Energy levels and eigenvectors at 10°.

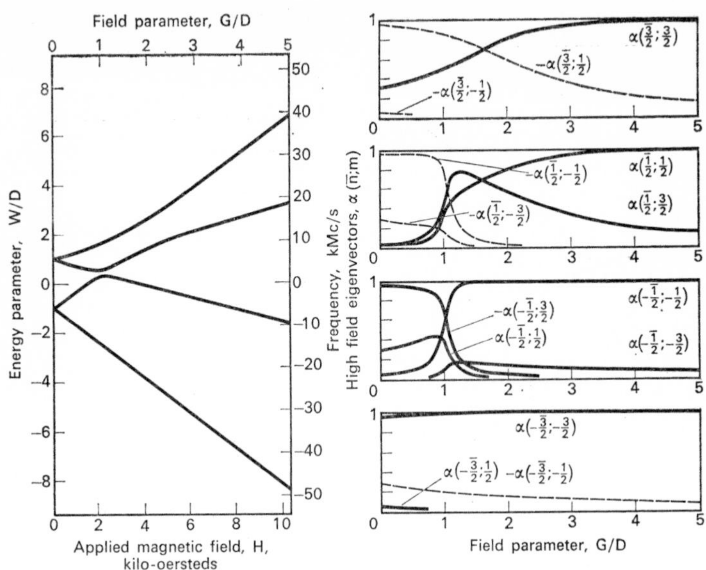

FIG. 3. Energy levels and eigenvectors at 20°.

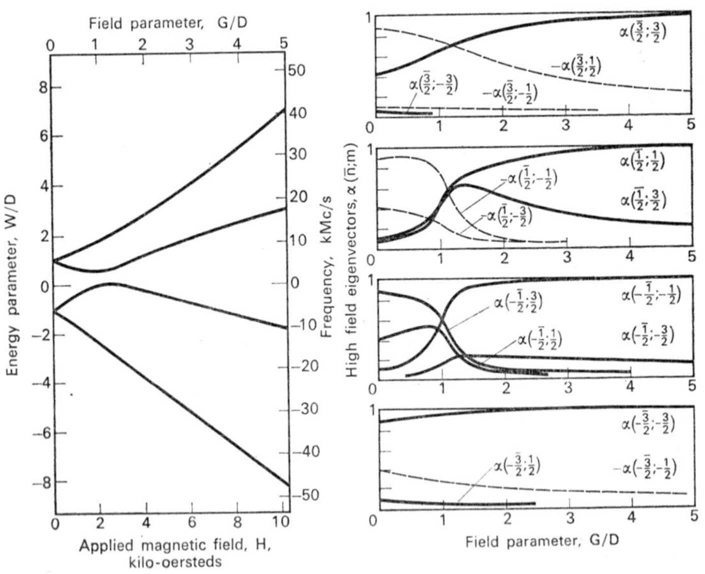

FIG. 4. Energy levels and eigenvectors at 30°.

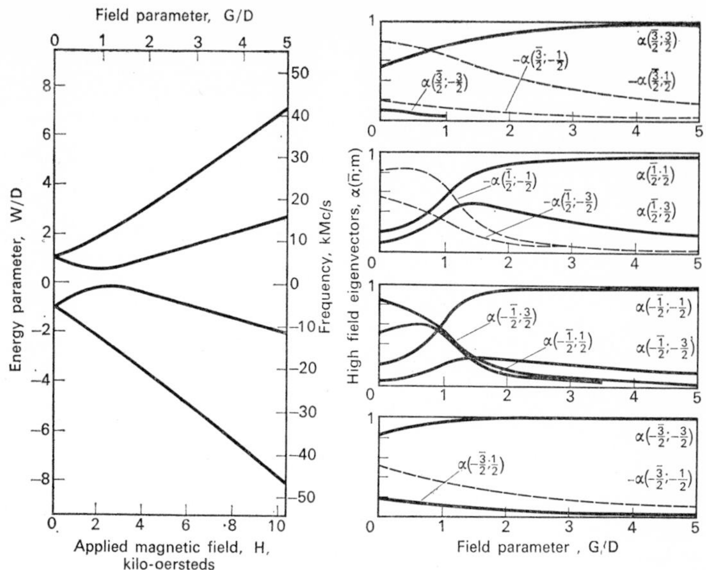

Fig 5. Energy levels and eigenvectors at 40°.

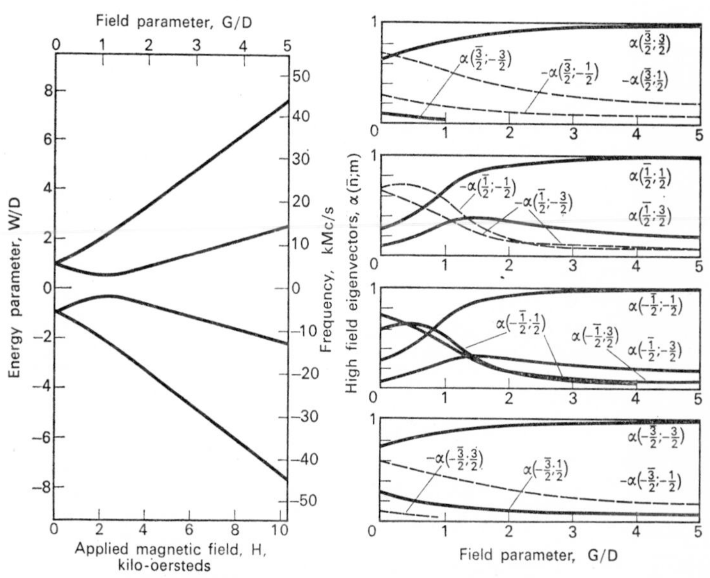

Fig. 6. Energy levels and eigenvectors at 50°.

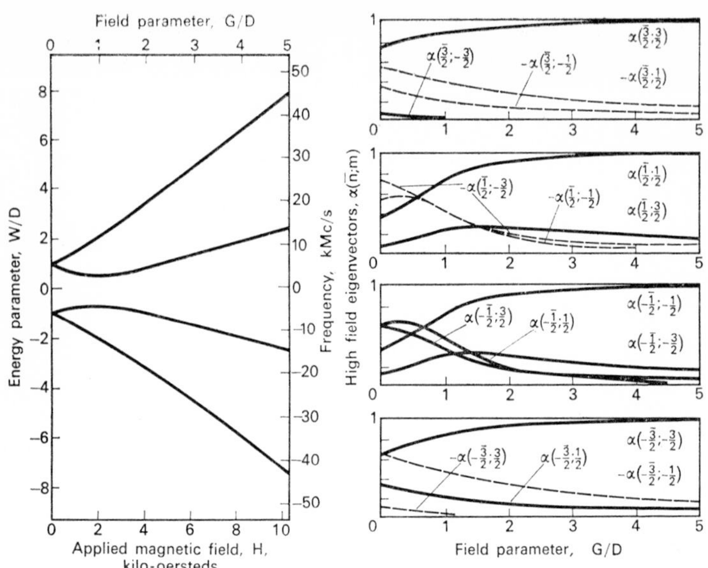

FIG. 7. Energy levels and eigenvectors at 54.7°.

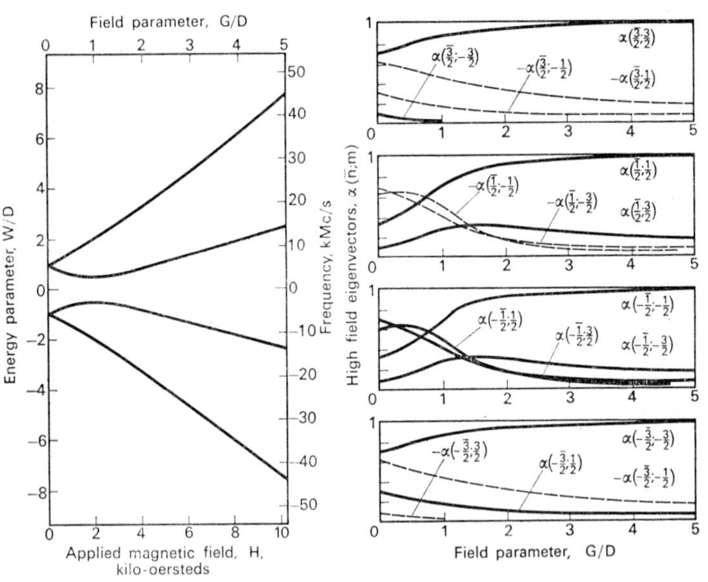

FIG. 8. Energy levels and eigenvectors at 60°

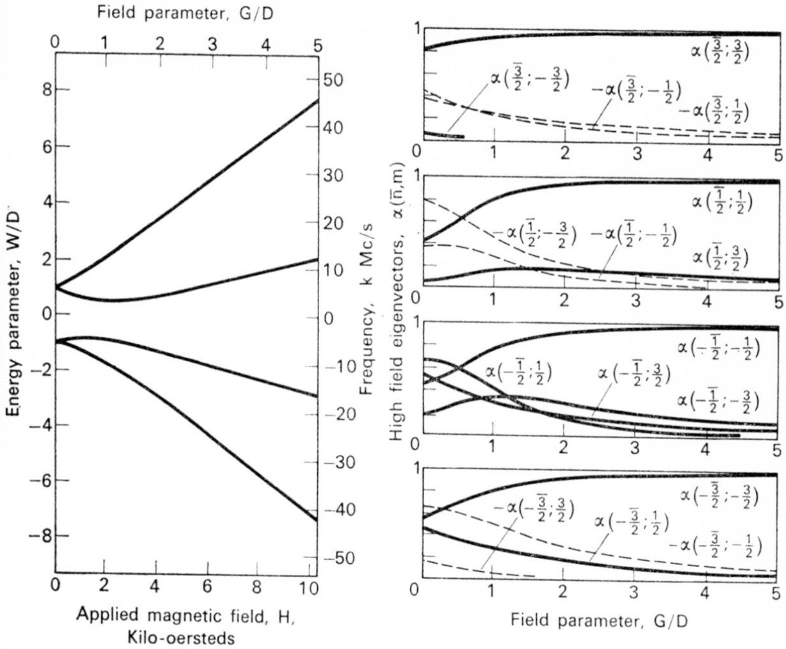

Field parameter, G/D
Energy parameter, W/D
Frequency, k Mc/s
Applied magnetic field, H, Kilo-oersteds
High field eigenvectors, α(n̄,m)
Field parameter, G/D

FIG. 9. Energy levels and eigenvectors at 70°.

Field parameter, G/D
Energy parameter, W/D
Frequency, kMc/s
Applied magnetic field, H, kilo-oersteds
High field eigenvectors, α(n̄;m)
Field parameter, G/D

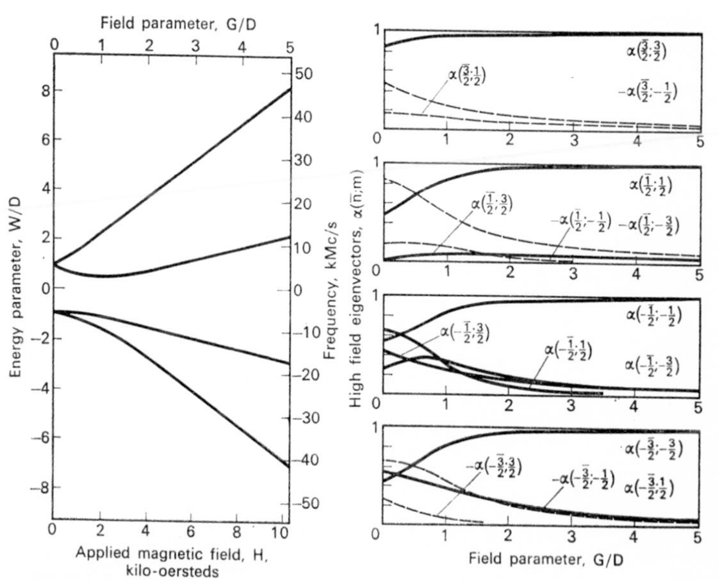

FIG. 10. Energy levels and eigenvectors at 80°.

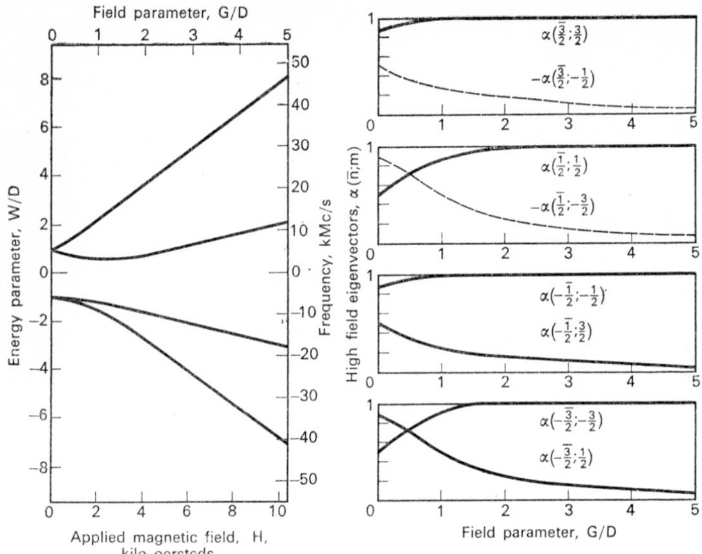

FIG. 11. Energy levels and eigenvectors at 90°.

splitting $2D$. Similar energy level diagrams were computed by P. M. Parker[11] for the case of nuclear spin resonance with nuclear quadrupole splitting present which is described by the same type of Hamiltonian.

As a convenient way to identify the energy levels W, a quantum number $\bar{n}$ ranging from $-\bar{\frac{3}{2}}$ to $+\bar{\frac{3}{2}}$ is used *in order of increasing energy*. Thus $W(-\bar{\frac{3}{2}})$ is the lowest, $W(\bar{\frac{3}{2}})$ the highest energy level. It is easily shown that, for all angles θ and $x' = 0$, $y(-\bar{\frac{3}{2}}) = y(-\bar{\frac{1}{2}}) = -1$ and $y(\bar{\frac{1}{2}}) = y(\bar{\frac{3}{2}}) = 1$. As a matter of mathematical curiosity it may be mentioned that, irrespective of θ at $x = 1$, $y(\bar{\frac{1}{2}}) = 1/2$.

The eigenstates $|\,\bar{n}\,\rangle$ (using Dirac's "ket" notation) associated with energy levels $W(\bar{n})$ can be expanded in the form

$$|\,\bar{n}\,\rangle = \sum_{m=-3/2}^{3/2} \alpha(\bar{n};\, m)\,|\,m\,\rangle. \tag{5}$$

Here, $|\,m\,\rangle$ are eigenstates of a Zeeman Hamiltonian $\mathscr{H} = g\beta HS_z$. The $\alpha(\bar{n};\, m)$ are amplitudes of eigenvector components or, more briefly, eigenvectors and form a normalized and orthogonal system of coefficients. With high applied magnetic field H, $|\bar{n}\rangle \rightarrow |n\rangle$ and $\alpha(\bar{n}; m) \rightarrow 1$; therefore, $|m\rangle$ are termed high-field eigenstates and $\alpha(\bar{n}; m)$ high-field eigenvectors.

Eigenvectors $\alpha(\bar{n}; m)$ are obtained as solutions of linear homogeneous equation systems, the matrix of which forms the secular equation (4) with the particular eigenvalue $W(\bar{n})$ inserted. Since this matrix depends on φ, the $\alpha(\bar{n}; m)$ are also functions of φ. The computations were carried out for $\varphi = 0$, with θ restricted to $0 < \theta < \pi/2$ and negative sign of $D = -D'$.

This choice implies that the crystalline axis lies in the positive quadrant of the x-z plane and it results in real eigenvectors $\alpha(\bar{n}; m)$. These are plotted in the right-hand sections of Figs. 1 through 11, adjacent to plots of the corresponding eigenvalues $W(\bar{n})$. Negative $\alpha(\bar{n}; m)$ are indicated by dashed lines.

A nonzero φ would, in general, result in new complex eigenvectors $\alpha'(\bar{n}; m) = [\exp i(m-\bar{n})\varphi]\ \alpha(\bar{n}; m)$. Taking $\pi/2 < \theta < \pi$ or $\varphi = \pi$ would change the sign of every second eigenvector, that is, of those with $m = n \pm 1$ and $m = n \pm 3$. The same is true for a change of sign of D, but then, in addition, every $\bar{n}$ and m and energy eigenvalue has to be replaced by its negative. It is obvious that such transformations do not change the physical situation as far as transition probabilities are concerned.

IV. Transition Probabilities

There are several ways in which transition probabilities could be evaluated and plotted. One way would be to consider transitions induced by radiation of given polarization. With eigenvectors belonging to the Hamiltonian (3), the obvious RF magnetic field polarizations to consider are those with RF H-field linear and parallel to, or circular and perpendicular to, the applied field. But transitions due to any other polarization could be evaluated as well. Perhaps more natural from a theoretical point of view would be an evaluation of the maximum transition probability. This requires a particular—in general—elliptical, polarization for excitation, which of course should be evaluated too. All polarizations orthogonal to this (which in general are elliptical as well), and which describe a plane in space having complex components, are associated with zero transition probability. Taking into account these different viewpoints and the six transitions which are possible between four energy levels, it appears that an unrealistically high number of graphs would be necessary to describe the transition probabilities properly. Furthermore, in maser design it is usually sufficient to know the order of magnitude of transition probabilities of particular lines, because often other factors may be more important. Therefore, no plots of transition probabilities are presented. On the other hand, enough of the pertinent formalism is given below so that any transition probability can be evaluated from the eigenvectors plotted.

Following essentially Bloembergen, Purcell and Pound,[12] with slight generalization, the transition probability w describing the rate of transitions per ion from a lower state $\bar{n}$ to a higher state $\bar{n}' > \bar{n}$ is given by

$$w_{\bar{n}\to\bar{n}'} = \frac{1}{4}\left(\frac{2\pi g\beta H_1}{h}\right)^2 g(\nu-\nu_0)\,|\,\langle\bar{n}'\,|\,S_1\,|\,\bar{n}\rangle\,|^2. \tag{6}$$

Here H_1 is the amplitude of the exciting RF magnetic field, $g(\nu-\nu_0)$ is a normalized function describing the line shape $\int g(\nu-\nu_0)\,d\nu = 1$, and S_1 is a spin operator reflecting the polarization of the inducing RF magnetic field. If the RF magnetic field is described by the real parts of $H_x = H_1 a e^{i\omega t}$, $H_y = H_1 b e^{i\omega t}$, $H_z = H_1 c e^{i\omega t}$ with "complex direction cosines" a, b, c accounting for elliptical polarization,

$$a^*a + b^*b + c^*c = 1,$$

then

$$S_1 = a^*S_x + b^*S_y + c^*S_z. \tag{7}$$

Matrix elements for S_1 occurring squared in (6) are linear combinations of the following three:

$$\langle\bar{n}'\,|\,S_z\,|\,\bar{n}\rangle = \sum_{m=-3/2}^{+3/2} m\alpha(\bar{n}';\,m)\alpha(\bar{n};\,m), \tag{8}$$

$$\langle \bar{n}' \mid S_+ \mid \bar{n} \rangle = \sum_{m=-3/2}^{+1/2} [S(S+1)-(m+1)m]^{1/2} \alpha(\bar{n}'; m+1)\alpha(\bar{n}; m), \tag{9}$$

$$\langle \bar{n}' \mid S_- \mid \bar{n} \rangle = \sum_{m=-1/2}^{+3/2} [S(S+1)-(m-1)m]^{1/2} \alpha(\bar{n}'; m-1)\alpha(\bar{n}; m). \tag{10}$$

The square root in (9) and (10) takes on the values $\sqrt{3}$, 2 and $\sqrt{3}$. For example, with linear polarization in the z-direction, $H_z = H_1 \cos \omega t$ and $S_1 = S_z$. For circular polarization perpendicular to the z-direction, $H_x = (1/\sqrt{2})H_1 \cos \omega t$, $H_y = \pm(1/\sqrt{2})H_1 \sin \omega t$ and $S_1 = (1/\sqrt{2})S_{\pm}$. For linear polarization in the x direction, $H_x = H_1 \cos \omega t$ and $S_1 = S_x = \frac{1}{2}(S_+ + S_-)$. Similarly, in the y direction $H_y = H_1 \cos \omega t$ and $S_1 = S_y = (1/2i)(S_+ + S_-)$.

The expression (7), or more correctly, the associated matrix element, can be interpreted as a scalar product of (a^*, b^*, c^*) with $\langle \bar{n}' \mid S \mid \bar{n} \rangle$. It should be noted that, in general, all components can be complex. As a consequence of this interpretation, the maximum transition probability occurs if H_{rf} or (a, b, c) is parallel in space and conjugate complex in phase to $\langle \bar{n}' \mid S \mid \bar{n} \rangle$. Since for real eigenvectors the matrices (8), (9), (10) are all real, it follows that $\langle \bar{n}' \mid S_x \mid \bar{n} \rangle$ and $\langle \bar{n}' \mid S_z \mid \bar{n} \rangle$ are real, whereas $\langle \bar{n}' \mid S_y \mid \bar{n} \rangle$ is imaginary. Thus, for all ruby lines, the polarization for maximum transition probability will be a linear combination of H_x and H_z components with an H_y component in quadrature. In a similar fashion, a set of complex direction cosines can be found which causes the scalar product of (a^*, b^*, c^*) with $\langle \bar{n}' \mid S \mid \bar{n} \rangle$, and hence the transition probability, to vanish. These vectors (a, b, c) describe a plane orthogonal to the vector for maximum transition probability.

It should be noted that frequently the complete formula (6) is not used to evaluate and compare transition probabilities. Instead, usually only the squared matrix element $|\langle \bar{n}' | S_1 | \bar{n} \rangle|^2$ is computed and this is then compared with a simple standard transition. The obvious standard is the transition $-1/2 \rightarrow +1/2$ of an $S = 1/2$ Zeeman doublet induced by circular polarization. This is described, in our notation, by $|\langle +1/2 | (1/\sqrt{2})S_+ | -1/2 \rangle|^2 = 1/2$. Accordingly, transitions involving a squared matrix element of order 1 or greater are considered strong, while perhaps 1/100 is typical of weak transitions.

V. Special Cases

5.1. $\theta = 0°$

The energy levels are parts of straight lines $y = 1 \pm \frac{1}{2}x$, $-1 \pm \frac{3}{2}x$ with change of slope for some of them at $x = 1$ and 2. Eigenvectors are ± 1 and 0 only, again joined for some levels at $x = 1$ and 2. The minus sign of eigenvectors at 0° has no significance; it is only used to preserve continuity to neighboring angles.

At $\theta = 0°$ and $x < 2$, the labeling of energy levels by high field quantum numbers *in order of increasing energy* is perhaps not the usual one. In this paper, however, it seems appropriate because, with this terminology, in going from $\theta = 0°$ to other orientations, the notation of states stays the same. It may be pointed out that energy levels defined in this fashion should be considered as continuous functions of applied field without cross-overs (see Fig. 1). The reason is that any off-diagonal perturbation will indeed prevent levels from intercepting by perturbation theory arguments.

Only three transitions are allowed:

$$0 < x < 1: \langle +\tfrac{3}{2} | S_+ | +\tfrac{1}{2} \rangle^2 = 4$$

$$\langle +\tfrac{3}{2} | S_- | -\tfrac{1}{2} \rangle^2 = \langle +\tfrac{1}{2} | S_+ | -\tfrac{3}{2} \rangle^2 = 3,$$

$$1 < x < 2: \langle +\tfrac{3}{2} | S_+ | -\tfrac{1}{2} \rangle^2 = 4$$

$$\langle +\tfrac{3}{2} | S_- | +\tfrac{1}{2} \rangle^2 = \langle +\tfrac{1}{2} | S_+ | -\tfrac{3}{2} \rangle^2 = 3,$$

$$2 < x: \quad \langle +\tfrac{1}{2} | S_+ | -\tfrac{1}{2} \rangle^2 = 4$$

$$\langle +\tfrac{3}{2} | S_+ | +\tfrac{1}{2} \rangle^2 = \langle -\tfrac{1}{2} | S_+ | -\tfrac{3}{2} \rangle^2 = 3.$$

It is interesting to note that, for $0 < x < 2$, one transition requires opposite polarization from the others. This was verified in an experiment. Resonance absorption was measured for this and another transition in a propagating comb-type slow-wave structure having regions of predominantly circular polarization. Reversal of applied magnetic field results in drastic increase of one and reduction of the other line.

5.2. $\theta = 54.74°$, $\cos^2 \theta = 1/3$

For this angle, the fourth-order secular equation reduces to a biquadratic one. The four eigenvalues are $y = \pm[1+\frac{5}{4}x^2 \pm (3x^2+x^4)^{1/2}]^{1/2}$. This implies an up-down symmetry $y(-\bar{n}) = -y(\bar{n})$. The closest approach of the two middle eigenvalues is $y(+\bar{\frac{1}{2}}) - y(-\bar{\frac{1}{2}}) = 1$ at $x = 1$. A similar symmetry relation holds for eigenvectors $\alpha(-\bar{n};\ -m) = (\bar{n}m/\mid \bar{n}m \mid)\alpha(\bar{n}; m)$. As a consequence, some transition probabilities for linear polarization are identical, namely

$$\langle -\bar{\tfrac{1}{2}} \mid S_z \mid -\bar{\tfrac{3}{2}} \rangle = \langle +\bar{\tfrac{3}{2}} \mid S_z \mid +\bar{\tfrac{1}{2}} \rangle$$

and

$$\langle +\bar{\tfrac{1}{2}} \mid S_z \mid -\bar{\tfrac{3}{2}} \rangle = -\langle +\bar{\tfrac{3}{2}} \mid S_z \mid -\bar{\tfrac{1}{2}} \rangle .$$

The analogous is not true for other polarizations.

5.3. $\theta = 90°$

The secular equation can be factorized into two quadratic equations with the solutions

$$y(\bar{\tfrac{3}{2}}) = \frac{x}{2}+(1+x+x^2)^{1/2},$$

$$y(\bar{\tfrac{1}{2}}) = -\frac{x}{2}+(1-x+x^2)^{1/2},$$

$$y(-\bar{\tfrac{1}{2}}) = \frac{x}{2}-(1+x+x^2)^{1/2},$$

$$y(-\bar{\tfrac{3}{2}}) = -\frac{x}{2}-(1-x+x^2)^{1/2}.$$

Each state contains only two eigenvectors, namely $\alpha(\bar{n}; n)$ and $\alpha(\bar{n}; n\pm 2)$. In addition, $\alpha(\bar{n}; n) = \alpha(\overline{n\pm 2}; \bar{n}\pm 2)$ and $\alpha(\bar{n}; n\pm 2) = -\alpha(\overline{n\pm 2}; n)$. As a result, transition probabilities between adjacent levels $\bar{n} \rightarrow \overline{n+1}$ contain only matrix elements of S_+ and S_-, the same being true for $-\frac{\bar{3}}{2} \rightarrow +\frac{\bar{3}}{2}$. Double jumps $\bar{n} \rightarrow \overline{n+2}$ are described by nonvanishing elements of S_z only.

VI. Paramagnetic Resonance Spectra

In Figs. 12 through 17 some resonance spectra are shown for signal frequencies of 5.18, 6.08, 9.30, 12.33, 18.2 and 23.9 kmc. The plots show resonance fields as functions of the angle between crystalline axis and applied field. Measurements have been carried out at all of these frequencies to varying extents, although measured values are recorded only on Figs. 14 and 15. Generally, these spectra have been used in the laboratory to align ruby crystals by

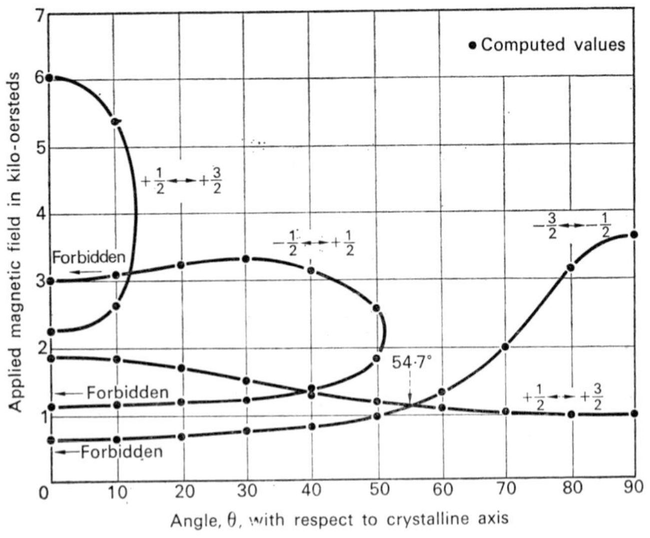

FIG. 12. Paramagnetic resonance spectrum of Cr^{+++} ions in ruby at signal frequency 5.18 kmc.

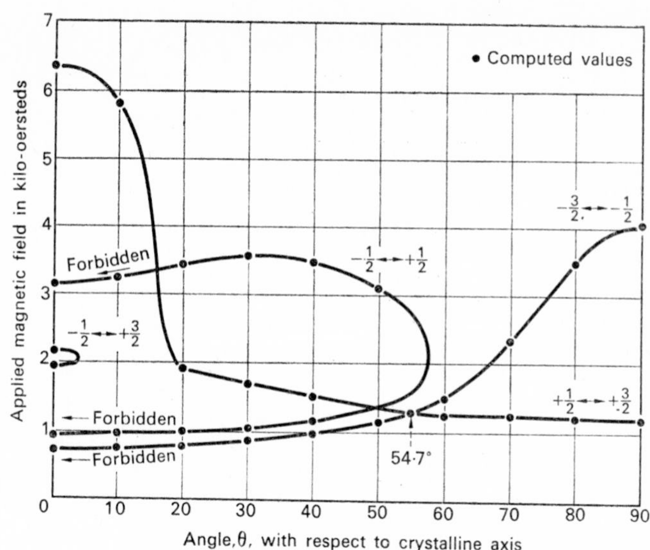

FIG. 13. Resonance spectrum at 6.08 kmc.

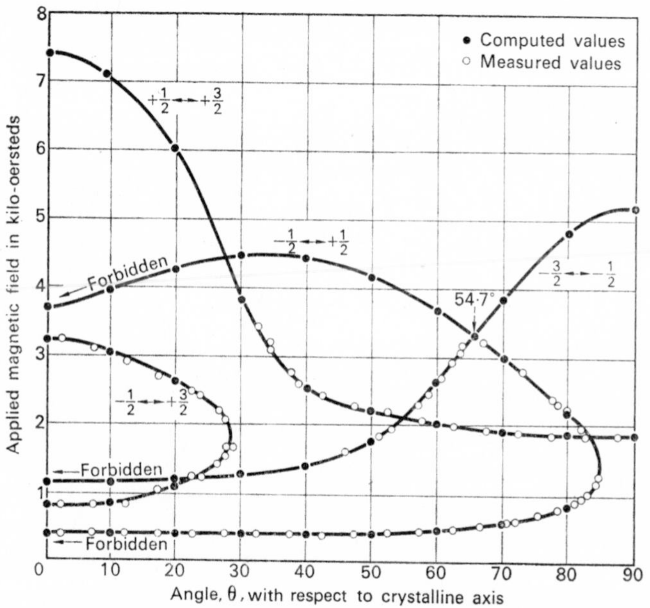

FIG. 14. Resonance spectrum at 9.30 kmc.

F

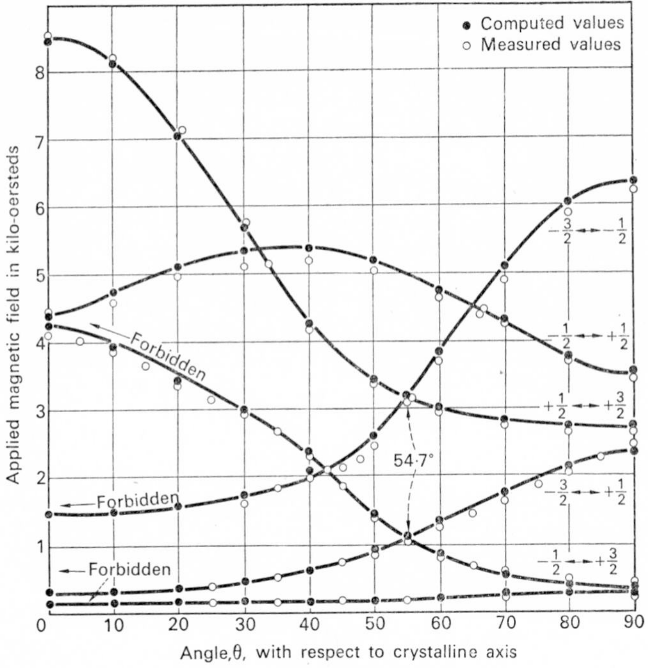

FIG. 15. Resonance spectrum at 12.33 kmc.

resonance for maser experiments. They have proved accurate to about ± 50 gauss.

Measurements at 9.3 kmc are an extension of Geusic's work[(4)] and confirm his results. Results at 12.33 kmc show some discrepancy between theory and experiment, which, however, is believed to be caused by inadequate magnetic field measuring equipment used in an experiment designed for other purposes. As a general rule, the spectra show two looping lines if $\nu < 2D$. Lines marked "forbidden" are strictly forbidden at 0° only. Usually, however, they can be followed quite close to 0° by use of more sensitivity in the spectrometer. An exception is the line shown on the graphs having the lowest resonance field at 0° if

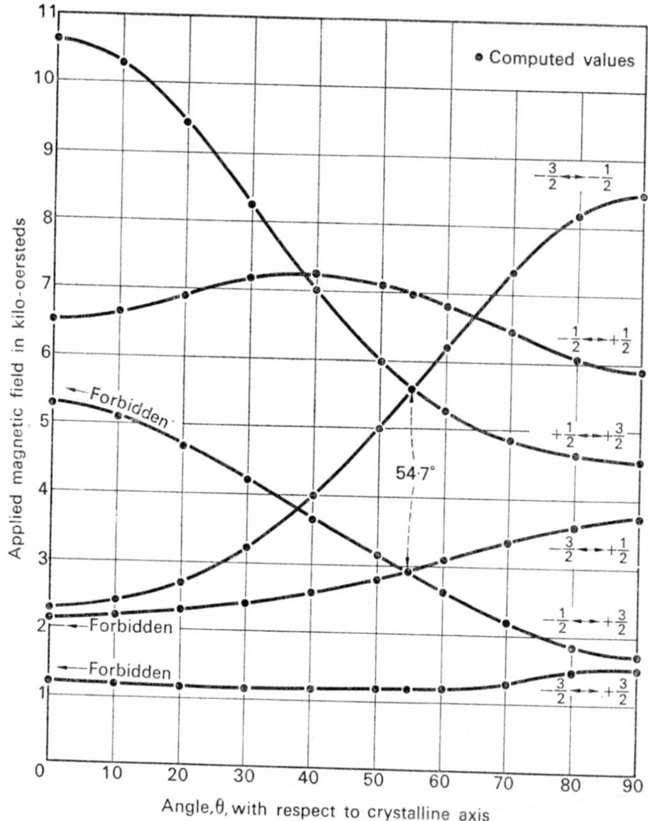

FIG. 16. Resonance spectrum at 18.2 kmc.

$\nu < \frac{6}{5}D'$. It has the second lowest resonance field if $\frac{6}{5}D' < \nu < \frac{3}{2}D'$ and the third lowest if $\frac{3}{2}D' < \nu < 3D'$. It originates between $-\frac{3}{2}$ and $+\frac{3}{2}$ eigenstates at 0° and is more strongly forbidden than the other forbidden lines; hence it usually ceases to be measurable at about 30°.

For reasons of symmetry, all lines approach 0° and 90° with zero slope $dH/d\theta$. Experimentally, it has been found that most lines are rather narrow at 90° and similarly at 0°, whereas they broaden in proportion with $dH/d\theta$. This behavior is expected from crystalline imperfections if these can be interpreted as fluctuations throughout the crystal of the direction of the crystalline axis.

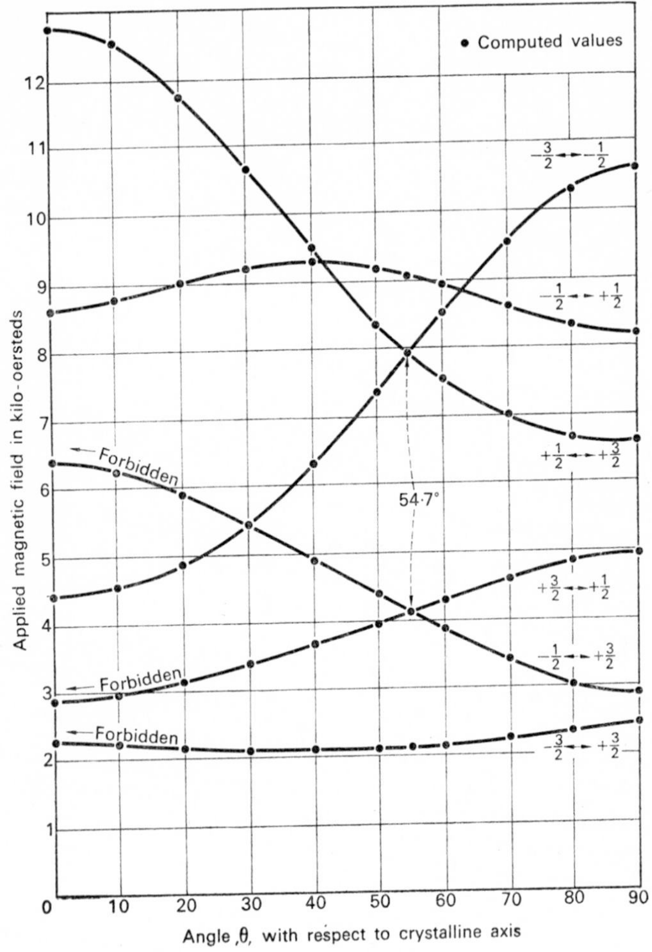

FIG. 17. Resonance spectrum at 23.9 kmc.

VII. Acknowledgment

The author wishes to thank many colleagues at Bell Telephone Laboratories for suggestions in the course of this study. He is particularly indebted to H. E. D. Scovil and J. E. Geusic. Miss M. C. Gray programmed and supervised the numerical calculations.

References

(1) Bloembergen, N., Proposal for a new-type solid-state Maser, *Phys. Rev.*, **104**, October 15, 1956, p. 324.

(2) Scovil, H. E. D., Feher, G. and Seidel, H., Operation of a solid-state Maser, *Phys. Rev.*, **105**, January 15, 1957, p. 672.

(3) Scovil, H. E. D., The three-level solid-state Maser, *Trans. I.R.E.*, **MTT-6**, January 1958, p. 29.

(4) Makhov, G., Kikuchi, C., Lambe, J. and Terhune, R. W., Maser action in Ruby, *Phys. Rev.*, **109**, February 15, 1958, p. 1399.

5) Bleaney, B. and Stevens, K. W. H., Paramagnetic resonance, *Rep. Prog. Phys.*, **16**, 1953, p. 107.

(6) Bowers, K. D. and Owen, J., Paramagnetic resonance II, *Rep. Prog. Phys.*, **18**, 1955, p. 304.

(7) Manenkov, A. A. and Prokhorov, A. M., *J. Exp. Theor. Phys.* (*U.S.S.R.*), **28**, 1955, p. 762.

(8) Geusic, J. E., *Phys. Rev.*, **102**, June 15, 1956, p. 1252; also Ph.D. dissertation, Ohio State Univ., 1958.

(9) Zaripov, M. and Shamonin, I., *J. Exp. Theor. Phys.* (*U.S.S.R.*), **30**, 1956, p. 291.

(10) Bruger, K., Ph.D. dissertation, Ohio State Univ., 1958.

(11) Parker, P. M., *J. Chem. Phys.*, **24**, 1956, p. 1096.

(12) Bloembergen, N., Purcell, E. M. and Pound, R. V., Relaxation effects in nuclear magnetic resonance absorption, *Phys. Rev.*, **73**, April 1, 1948, p. 679.

PAPER 8*

Spin–Lattice Relaxation Times in Ruby at 34·6 Gc/s

J. H. PACE, D. F. SAMPSON and J. S. THORP

Royal Radar Establishment, Malvern, Worcs.

MS. received 1st April 1960, in revised form 20th May 1960

Abstract

Measurements of spin–lattice relaxation time in ruby have been made at 34·6 Gc/s by a pulse saturation method at temperatures from 1·4° to 90°K. With weak concentrations the values for the first-order transitions (e.g. 22 msec at 4·2°K) are of the same order of magnitude as those reported at lower frequencies, and the variation of relaxation time with temperature is in fair agreement with theory. A successive increase in relaxation time with the order of the transition, pronounced at 1·4°K, decreases with increasing temperature, and at about 77°K a common value of about 44 μsec is obtained for all transitions. The main effects of increasing concentration are to reduce the relaxation time and alter its temperature dependence.

§ 1. Introduction

The importance of measurements of spin–lattice relaxation times T_1 in paramagnetic materials has grown recently because of the rapid developments in the maser field. Although several theories have been proposed (e.g. Van Vleck 1940, Al'tshuler 1956, Gill 1959, unpublished) the physical basis of the relaxation process is not clearly understood, and the formulation of a general theory applicable to a range of materials awaits the accumulation of more experimental data. Specific problems occur, however, which demand knowledge of relaxation times and their dependence on

* *Proc. Phys. Soc.* **76,** 697–704 (1960).

certain parameters. Two examples are the extension of maser techniques for amplification or oscillation to the short millimetre wavelength region (Foner, Momo and Mayer, 1959), and the operation of lower frequency masers at elevated temperatures (Ditchfield and Forrester, 1958) with increased bandwidth and power handling capacity. In the former case the magnitude and frequency dependence of the relaxation time are important, and in the latter the effects of temperature and concentration. This paper describes measurements of the spin–lattice relaxation time in synthetic ruby made at 34·6 Gc/s with the object of investigating these effects. The temperature and concentration ranges covered were 1·4° to 90°K and 0·03 % Cr to 0·78 % Cr respectively, using the fixed polar angle $\theta = 90°$. The concentration is defined as the ratio of weights of chromium and aluminium oxide. Ruby was selected because it is known to be a suitable material for many maser applications and the frequency was chosen, bearing in mind the current interest in millimetre wavelength masers, as being the highest at which sufficient microwave power and components were easily obtainable. Preliminary results have already been reported by the authors Pace, Sampson and Thorp (1960).

§ 2. Apparatus and Technique

2.1. *General Features*

A pulse saturation method was used in which the transition studied was saturated by a short pulse, recovery to thermal equilibrium being observed by measuring the absorption of a low power, c.w. monitoring signal of the same frequency (cf. Davis, Strandberg and Kyhl, 1958).

The general arrangement of the equipment followed conventional lines. The low temperature assembly however was a straight silver-plated cupro-nickel waveguide (inside dimensions 0·280 in. × 0·140 in., wall thickness 0·005 in.) terminated by a removable plane plunger on which the specimen was mounted. Use of this system removed difficulties of cavity tuning on cooling, permitted large samples to be used, and enabled the whole structure to be made easily removable for sample changing at helium temperatures. It also led to a small structure, a particularly valuable feature at high magnetic fields where space in the gap is restricted. Magnetic fields of up to 16 000 gauss over a 4·5 cm gap were provided by a Newport Instruments Type D electromagnet. These were uniform to at least 1 in 10^4 over a volume of about 2 cm cube, and field stabilities of the same order were readily obtained using a a bank of batteries to drive the electromagnet. Additional coils enabled low frequency field modulation to be supplied when required.

2.2. *Microwave System*

The microwave arrangement, shown in figure 1, consisted of a waveguide bridge, a low power monitor source, a high power pulsed source, and a super-heterodyne receiver.

One arm of the bridge was the low temperature assembly described above. Balance was obtained when off magnetic resonance

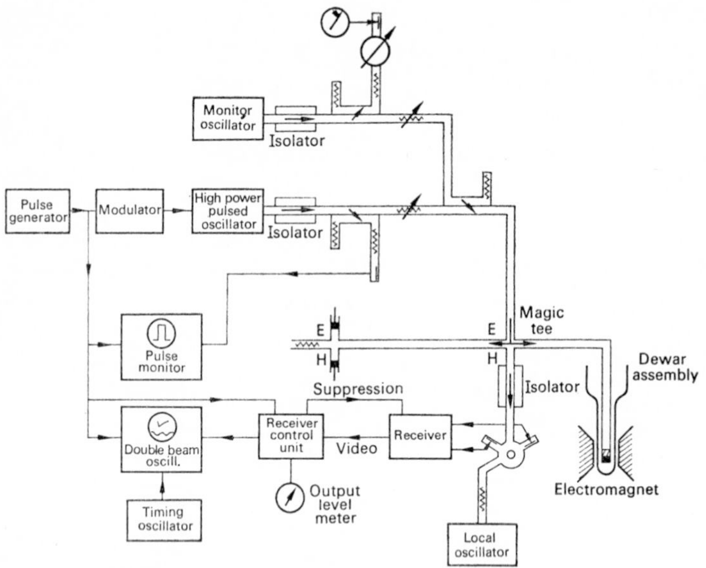

FIG. 1. Microwave arrangement for relaxation time measurement.

by adjusting the *E–H* tuner in the other arm for cancellation of reflected power into the balanced mixer. When absorption occurred in the sample, i.e. on magnetic resonance, the bridge became unbalanced and a signal proportional to the degree of absorption was fed into the receiver.

The bridge input consisted of two signals mixed in a 6 dB directive feed, viz: (i) a low power, c.w. monitoring signal, variable between 1 μw and 10 μw to allow for sensitivity changes, supplied from a VX.5023 reflex klystron and (ii) a high power pulse, generated by an Elliott Type B.579 floating drift tube klystron. The latter valve was pulsed by a hard valve modulator driven by a pulse generator, and produced pulses variable in length from 50 μsec to 500 μsec at pulse repetition frequencies up to 25 c/s. The maximum pulse power available was 6 watts peak. With a pulse duration of 200 μsec the minimum power required for saturation was 60 mW at 1·4°K.

The frequency of measurement was determined by the fixed frequency of the drift tube klystron, 34·6 kMc/s, to which the monitor oscillator was tuned. It was unnecessary to stabilize the monitor oscillator because with the samples used, whose line widths were not less than about 30 gauss, saturation occurred if the pulse frequency was within ± 50 Mc/s of the absorption frequency.

The receiver was of conventional design, employing a second VX.5023 as the local oscillator. An intermediate frequency bandwidth of 10 Mc/s allowed operation without frequency stabilization or automatic frequency control. The overall gain to the second detector was 110 dB at 45 Mc/s and the total noise factor, including the mixer, was 15 dB. The recovery time to maximum sensitivity, after suppression during the saturating pulse, was 5 μsec.

The second detector output was coupled via a cathode follower to one trace of a Cossor double beam oscilloscope type 1049, the other trace displaying a sinusoidal timing waveform. Use of d.c. amplification in the oscilloscope enabled the maximum and minimum levels of absorption to be displayed independent of the form and duration of the recovery curve. A facility for simultaneous interruption of the timing waveform and the drive pulse to the modulator enabled the maximum absorption level to be displayed independently. In this way a composite photographic record was taken showing (*a*) the recovery curve, (*b*) the maximum absorption level in the absence of the saturating pulse superimposed on this, and (*c*) the timing waveform (figure 2); the value

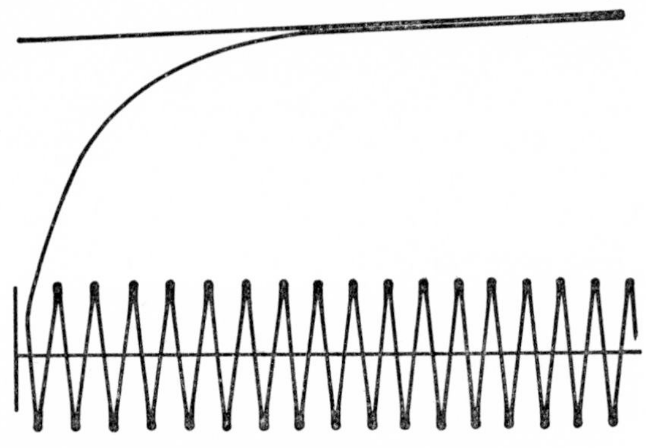

FIG. 2. Presentation of recovery curve.

of relaxation time could be derived from this. The overall accuracy of the measurements was estimated to be within about $\pm 10\%$, and the minimum measurable relaxation time, set by the receiver recovery, was about 10 μsec.

§ 3. Relaxation Times for Low Chromium Concentration

The primary factor governing the orientation chosen was the need to assess ruby as a material for pulsed field maser experiments. In these use of the maximum available relaxation time would reduce the complexity of the circuitry required. Since it was known from the work of Bloembergen *et al.* (1959) that cross relaxation processes can be very important and that they reduce relaxation times a position was selected at which these would be negligible. This orientation was with the *c* axis of the crystal perpendicular to the d.c. magnetic field, $\theta = 90°$ (cf. Mims 1959). The first

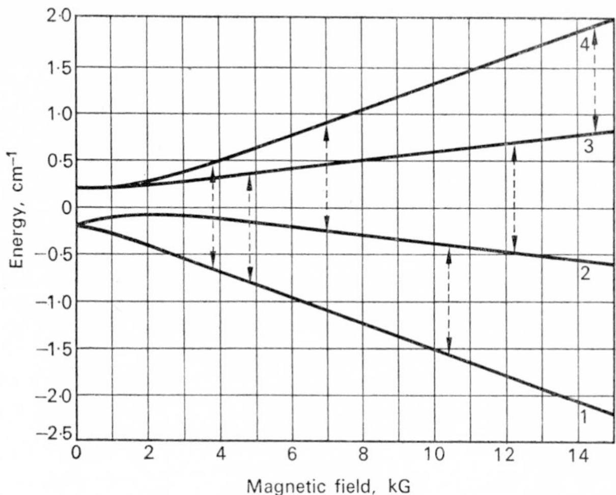

FIG. 3. Energy level diagram for ruby at $\theta = 90°$ showing 34·6 Gc/s transitions.

measurements were made on a sample of nominal concentration 0·1% Cr as it was known that crystals of this concentration had been successfully used in three level masers. (Analysis subsequently showed that this crystal contained 0·03% Cr). Figure 3 shows the energy level diagram at $\theta = 90°$ together with possible transitions at 34·6 Gc/s. The measured relaxation times are given in the table.

RELAXATION TIMES (msec) IN NOMINAL 0·1% Cr RUBY
34·6 Gc/s, $\theta = 90°$

Transition	1·4°K	4·2°K	10·1°K	20·3°K	56°K	77°K	90°K
3–4	60	21	16	4	0·086	—	0·016
2–3	64	16	10	8	0·083	0·045	0·015
1–2	59	22	10	6	0·10	0·044	0·017
2–4	147	54	12	10	0·16	0·049	—
1–3	100	56	—	12	0·13	—	—
1–4	296	—	—	—	0·12	—	—

The conclusions which may be drawn from these results will now be discussed.

3.1. *Variation of Relaxation Time with Frequency*

The measured value of T_1 at 4°K for the 1–2 transition is 22 msec. As far as comparison with previous measurements can be made this is of the same order as those reported at 7·2 Gc/s (Mims*) and 9·3 Gc/s (Gill and Harvey, 1958, unpublished) and 24 Gc/s (Kikuchi *et al.* 1959). It therefore appears that over this range the spin–lattice relaxation time does not vary appreciably with frequency. No general theoretical rule for the frequency dependence can be made, since this is determined by the material and the coupling mechanism. It may be noted however that a variation as $(\text{frequency})^{-2}$ or $(\text{frequency})^{-4}$, predicted by Van Vleck for chromium alum, is not found.

3.2. *Variation of Relaxation Time with Temperature*

The effect of temperature on the relaxation time of the 1–2 transition is shown in figure 4. In the range of temperature from 1·4° to 20°K the relaxation time is approximately inversely proportional to temperature. This behaviour is similar to that reported in ruby at 9·3 Gc/s (Harvey, 1958, unpublished) and in potassium chromicyanide in the helium range when cross relaxation effects are negligible (Shapiro and Bloembergen, 1959): it would be expected if relaxation were due to a direct exchange of energy between the spin system and lattice phonons. Between 20° and 56°K this relation breaks down, and above 56°K the rapid decrease in relaxation time with increasing temperature is consistent with the predominance of Raman effects; preliminary observations suggest that the variation is as $(\text{temperature})^{-5}$. This is not quite so rapid as the $(\text{temperature})^{-7}$ suggested by the Van Vleck model. For the other transitions a broadly similar temperature dependence was found.

* Paper given at Radio-frequency Spectroscopy Conference, Oxford, 1959.

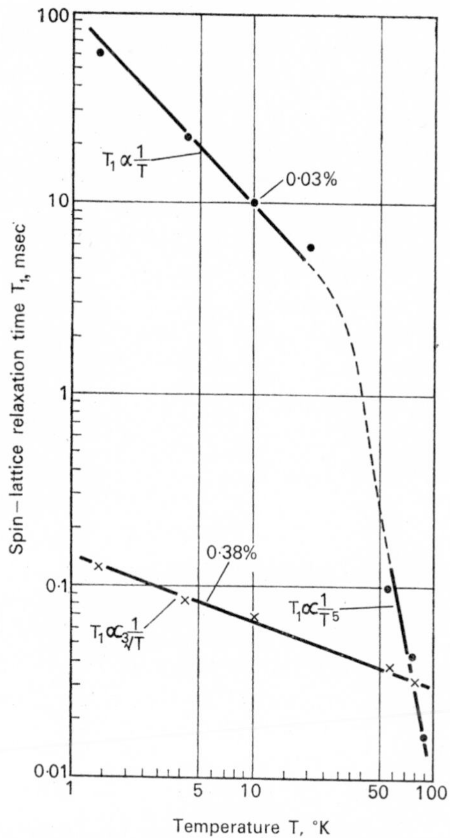

FIG. 4. Variation of relaxation time with temperature (1–2 transition, $\theta = 90°$).

3.3. *Relaxation Times for Transitions of Different Order*

At 1·4°K the relaxation times associated with the different transitions are not equal, and the second- and third-order transitions have significantly longer relaxation times than the first-order transitions although the recovery curves remain substantially single exponentials. This is contrary to predictions based on the Van Vleck model, but a recent development of Al'tschuler's theory (Gill, 1959, unpublished) has shown that different relaxation times for the various transitions are to be expected if the relaxation process is governed by near-neighbour interaction and gives close agreement with experiment. As the temperature is increased the difference between the relaxation times for the various transitions decreases, and a common value is reached at about 77°K. The occurrence of this common value has previously been observed in ruby at 77°K at 9·3 Gc/s (Gill, 1959, unpublished) and is presumably a consequence of the different relaxation process occurring at higher temperatures.

§ 4. Effects due to Increased Concentration

It is found that the relaxation time for a given transition is a function of the concentration of paramagnetic centres and the temperature. Ruby samples of known concentration were available only in small sizes, and measurements were therefore confined to the strong first-order transitions. The concentration range covered was from 0·02 % Cr to 0·38 % Cr.

4.1. *Behaviour at Fixed Temperatures*

At helium temperatures the relaxation time decreases with increasing concentration N. The results at 1·4°K are shown in figure 5 (lower curve) and can be expressed approximately by $T_1 \propto N^{-2}$. A similar concentration dependence has also been observed at 4·2°K and 9·3 Gc/s (de Grasse *et al.*, 1959). It may be noted that a concentration dependence is not predicted by theories based on the Van Vleck model, but is a consequence of theories based on near-neighbour interaction.

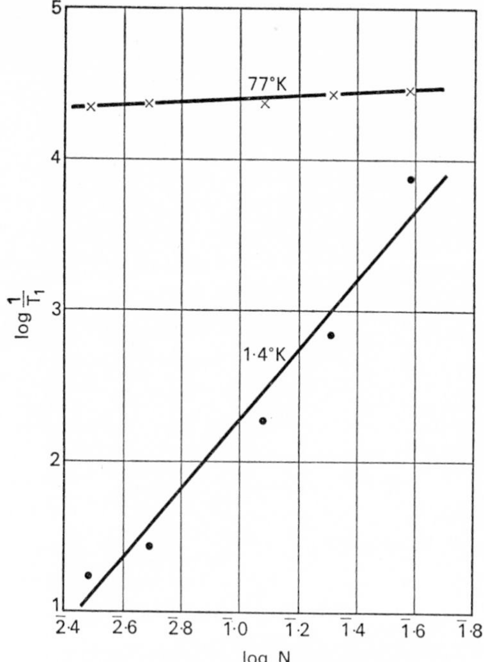

FIG. 5. Variation of relaxation time with concentration (1–2 transition, $\theta = 90°$).

At 77°K the effect of concentration is much smaller (figure 5, upper curve) and is approximately given by $T_1 \propto N^{-0 \cdot 1}$. This result is of importance in connection with the operation of three level masers at nitrogen temperatures, because it implies that a higher concentration can be used—thereby increasing both the inherent bandwidth and the power handling capacity of the maser—without undue increase in the pump power requirement.

4.2. *Relaxation Time–Temperature Dependence*

The dependence of relaxation time on temperature has already been discussed for a low concentration sample (figure 4, upper curve). The curve is characterized by two regions over which

$T_1 \propto T^{-1}$ and $T_1 \propto T^{-5}$ respectively. As the concentration is increased this behaviour changes and at concentrations above about 0·3 % Cr, a smooth variation from 1·4° to 90°K is found (figure 4, lower curve). It is found that, approximately, $T_1 \propto T^{-1/3}$ and it is thus clear that a different relaxation process occurs. This slow variation suggests that, with the correct material, maser action should be possible at relatively high temperatures.

It is interesting to note that somewhat similar concentration effects have been observed in other materials. In nickel fluosilicate Bowers and Mims (1959) found that in the helium range there was a general tendency for the relaxation times to be less temperature dependent at high concentrations while in potassium chromicyanide Shapiro and Bloembergen (1959) have shown that the smaller relaxation times observed at high concentrations can be explained by the growing importance of cross relaxation.

§ 5. Very High Concentration

One sample was available with a chromium concentration of 0·78 % Cr. In preliminary observations on this sample it was found that at 77°K a simple exponential recovery curve was obtained. This gave a relaxation time of about 30 μsec, as would be expected from the foregoing results. At helium temperatures, however, the recovery curve had a fast initial rise (corresponding to a relaxation time of about 15 μsec) followed by a much slower rise (corresponding to a 20 msec relaxation time). A similar effect at 9 Gc/s has been reported (de Grasse *et al.*, 1959) and the results suggest that the fast recovery is due to effects within the spin system while the slow recovery is at a rate comparable with the relaxation time for a dilute sample.

§ 6. Conclusions

Using a given polar angle, $\theta = 90°$, the spin–lattice relaxation time in ruby at 34·6 Gc/s has been shown to depend on the temperature, the transition, and the chromium concentration. Comparison with results obtained at lower frequencies suggest that the relaxation time is substantially independent of frequency. At low concentrations the temperature variation found agrees broadly with predictions based on the Van Vleck model; the remaining observations can best be explained by assuming that the dominant mechanism in the relaxation process is near-neighbour interaction.

Because of the long relaxation times obtainable ruby appears to be a promising material for pulsed field masers intended to amplify or oscillate at millimetre wavelengths. At lower microwave frequencies it appears that improvements are to be gained in the operation of masers at elevated temperatures by using ruby with high chromium concentration.

Acknowledgments

We wish to record our thanks to Dr. D. J. Howarth, Royal Radar Establishment, Malvern, for supplying the data given in figure 3, to the Chemical Inspectorate, Woolwich, for the analysis of the samples, and to colleagues in the Laboratory for helpful discussions.

This paper is published by permission of the Controller, Her Majesty's Stationery Office.

References

AL'TSHULER, S. A. (1956) *Bull. Acad. Sci. U.S.S.R.* (*Phys. Ser.*), **20,** (11), 1207.

BLOEMBERGEN, N., SHAPIRO, S., PERSHAN, P. S. and ARTMAN, J. D., (1959), *Phys. Rev.*, **114,** 445.

BOWERS, K. D. and MIMS, W. B. (1959) *Phys. Rev.*, **115,** 285.

DAVIS, C. F., STRANDBERG, M. W. P. and KYHL, R. L. (1958) *Phys. Rev.*, **111,** 1268.

DE GRASSE, R. W., GEUSIC, J. E., SHULZ-DU BOIS, E. O. and SCOVIL, H. E. D. (1959) *9th Interim Report on Microwave Solid State Devices*, Bell Telephone Laboratories.

DITCHFIELD, C. R. and FORRESTER, P. A. (1958) *Phys. Rev. Lett.*, **1,** 448.

FONER, S., MOMO, L. R. and MAYER, A. (1959) *Phys. Rev. Lett.*, **3,** 36.

KIKUCHI, C., LAMBE, J., MAKHOW, G. and TERHUNE, R. W. (1959) *J. Appl. Phys.*, **30,** 1061.

PACE, J. H., SAMPSON, D. F. and THORP, J. S. (1960) *Phys. Rev. Lett.*, **4,** 18.

SHAPIRO, S. and BLOEMBERGEN, N. (1959) *Phys. Rev.*, **116,** 1453.

VAN VLECK, J. H. (1940) *Phys. Rev.*, **57,** 426.

PAPER 9*

Spin–Lattice Relaxation and Cross-Relaxation Interactions in Chromium Corundum

A. A. MANENKOV and A. M. PROKHOROV

P. N. Lebedev Physics Institute, Academy of Sciences, U.S.S.R.

Submitted to *JETP* editor July 31, 1961

J. Exptl. Theoret. Phys. (*U.S.S.R.*) **42,** 75–83 (January, 1962)

Summary

Relaxation phenomena in chromium corundum ($Al_2O_3 \cdot Cr_2O_3$) single crystals were studied at liquid helium temperatures and a frequency of 9400 Mc by the pulse technique. The spin–lattice relaxation times were determined for samples with various Cr^{3+} concentrations. Spin–spin cross-relaxation interactions were discovered in the spectrum and were investigated for various splitting ratios between the Cr^{3+} ion energy levels. The times and amplitudes of the exponentials characterizing the cross-relaxation interactions were determined for a sample having a Cr^{3+} concentration of 0.15%. An interpretation of the results is presented.

1. Introduction

The establishment of thermal equilibrium between the spin system of paramagnetic ions and the lattice, as well as within the spin system, plays an important role in crystals that are used to achieve negative temperatures in the energy level system. In view of this there has been a marked growth of interest in recent years in the investigation of relaxation phenomena in paramagnetic crystals, particularly at low temperatures. Several investigations of relaxation in crystals have found practical application in paramagnetic quantum amplifiers. This research, stimulated by the

* *J. Exp. Theoret. Phys.* (*U.S.S.R.*) **15,** 54–59 (1962).

practical considerations of quantum electronics, has, in its turn, led to the development of the theory of paramagnetic relaxation. In particular, Bloembergen and others[1] have developed a theory of spin–spin cross-relaxation, which, as confirmed by several experiments, plays an extremely important role in the processes that establish thermal equilibrium with a spin system with many energy levels.

Kochelaev[2] and Anderson[3] have undertaken also to develop theories of spin–lattice relaxation. Although some of the peculiarities of paramagnetic relaxation at low temperatures have been qualitatively explained on the basis of cross-relaxation, a more detailed investigation of both cross-relaxation interactions and spin–lattice relaxation in paramagnetic crystals is needed.

The object of the present work was the detailed study of the relaxation phenomena in the spectrum of Cr^{3+} in monocrystals of corundum (Al_2O_3) at liquid helium temperatures. The relaxation of Cr^{3+} in Al_2O_3 has been the subject of many previous studies.[4–10] In our first paper,[4] devoted to the investigation of spin–lattice relaxation of Cr^{3+} in Al_2O_3, the continuous saturation method was used. This method is an indirect one for the determination of the spin–lattice relaxation time T_1, for to calculate T_1 from the saturation parameters determined directly from the experiments, it is necessary to know the spin–spin relaxation time T_2 and the transition probability for the observed paramagnetic resonance line, and also the intensity of the high-frequency field causing the saturation. Besides, it is difficult to separate the effects of spin–lattice and spin–spin cross-relaxation in this method.

In the present more detailed investigation of spin–lattice and cross-relaxation we have therefore used the pulse method of saturation of the paramagnetic resonance line. The pulse method is a direct one for measuring the relaxation times and permits the separation of spin–lattice and cross-relaxation effects, through the use of different pulse lengths. Because of this, it is possible to determine the spin–lattice relaxation and cross-relaxation times T_1 and T_{12} and to separate the dependence of T_1 and T_{12} on the paramagnetic ion concentration and temperature.

2. Method and Experimental Conditions

The experiments to study the relaxation of Cr^{3+} in Al_2O_3 were carried out with a superheterodyne spectrometer at 9400 Mc and liquid helium temperatures. The pulse method described earlier[11] was used. Saturation of the paramagnetic resonance line was produced by pulses varying in length from 0·8 millisec to 1 sec. Monocrystals of corundum with the following concentrations of Cr^{3+} ion relative to the diamagnetic Al^{3+} ions were studied: 0.05, 0.1, 0.15, 0.4, and 0.65%. The sample with 0.15% concentration was studied in much more detail because a strong manifestation of the cross-relaxation effect was observed in the region of concentrations from 0.1 to 0.15%.

3. Theoretical Treatment of Cross-Relaxation between Cr^{3+} Levels

In the lowest ground state of a Cr^{3+} ion situated in the crystalline field of the Al_2O_3 lattice and an external magnetic field there are four spin energy levels, corresponding to the electronic spin $S = \frac{3}{2}$. The rigorous analysis of the relaxation processes in such a system, taking into account the relaxation transitions evoked by the spin–lattice and spin–spin interactions, is quite complicated. However, certain rules concerning cross-relaxation in a system of four levels can be deduced by means of a simplified scheme. In this scheme we regard the system as made up of two pairs of levels E_1, E_2 and E_3, E_4, between which a spin–spin cross-relaxation is established, but the spin–lattice transitions proceed only within each pair of levels; for simplicity we assume that these transitions occur at the same rate (identical spin–lattice relaxation times for both pairs of levels).

Thus, in this simplified scheme we are neglecting the spin–lattice transitions between the levels 1 and 3, 1 and 4, 2 and 3, and 2 and 4. In the real system of levels of Cr^{3+} such neglect is valid when the probabilities of the indicated spin–lattice transitions is much less than the probabilities of the cross-relaxation transitions between the levels 1 and 2 or 3 and 4. If the splittings between the levels have a whole-number ratio $E_4-E_3 = m(E_2-E_1)$, $m = 1$, 2, 3, . . ., then the cross-relaxation between the level pairs E_1, E_2 and E_3, E_4 can occur on account of the simultaneous transitions of two (when $m = 1$) or more (when $m = 2, 3$) ions between which a spin–spin interaction exists. The total energy of the spin system is not changed by such cross-relaxation transitions.

Figure 1 shows a level scheme in which the splitting ratio is $m = 2$. In this system cross-relaxation can take place in the following way: a change in the orientation of the spin of one ion (transition from level 3 to level 4) is accompanied by a simultaneous change of orientation of the spins of two neighbouring ions (transitions from level 2 to level 1). For the population differences we have the following kinetic equations:

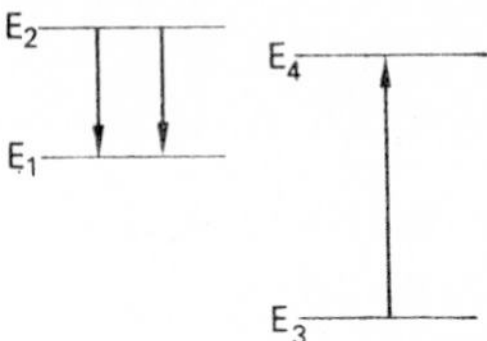

FIG. 1. Cross-relaxation transitions in a four-level system.

$$\frac{d(n_1-n_2)}{dt} = -\frac{1}{T_1}(n_1-n_2-n_1^0+n_2^0)+w_{cr}(n_2^m n_3-n_1^m n_4)$$

$$-2W(n_1-n_2),$$

$$\frac{d(n_3-n_4)}{dt} = -\frac{1}{T_1}(n_3-n_4-n_3^0+n_4^0)-\frac{w_{cr}}{m}(n_2^m n_3-n_1^m n_4), \qquad (1)$$

where n $(i = 1, 2, 3, 4)$ is the population of the energy level at thermal equilibrium, T_1 is the spin–lattice relaxation time, w_{cr} is the probability of cross-relaxation, and W is the probability of the $1 \rightarrow 2$ transition induced by the external high-frequency field.

When the separation between the levels is less than kT, Eq. (1) can be rewritten in the form

$$\frac{d\Delta n_{12}}{dt} = -\frac{1}{T_1}(\Delta n_{12}-\Delta n_{12}^0)+w'_{cr}(\Delta n_{34}-\beta\Delta n_{12})-2W\Delta n_{12},$$

$$\frac{d\Delta n_{34}}{dt} = -\frac{1}{T_1}(\Delta n_{34}-\Delta n_{34}^0)-\frac{w'_{cr}}{m}(\Delta n_{34}-\beta\Delta n_{12}), \tag{2}$$

where

$$\Delta n_{ij} = n_i-n_j, \quad w'_{cr} = w_{cr}n_2^m \approx \text{const}, \quad \beta = \Delta n_{34}^0/\Delta n_{12}^0.$$

When the probability of transitions induced by the external field is much less than the probabilities of spin–lattice and cross-relaxation transitions ($W \to 0$), the solution of (2) has the form:

$$\Delta n_{12} = A\exp(\alpha_1 t)+B\exp(\alpha_2 t)+\Delta n_{12}^0,$$

$$\Delta n_{34} = A\exp(\alpha_1 t)-\frac{B}{m}\exp(\alpha_2 t)+\Delta n_{34}^0, \tag{3}$$

where A and B are the amplitudes of the two exponentials, determined from the initial conditions

$$\alpha_1 = -1/T_1, \quad \alpha_2 = -1/T_1-1/T_{12},$$

$$T_{12} = m/(1+\beta m)w'_{cr}.$$

Usually $T_{12} \ll T_1$. Hence, in Eq. (3) it is the exponential with the smaller time constant α_2 that basically characterizes the cross-relaxation. The exponential with the time constant α_1 characterizes the spin–lattice relaxation.

Let us consider a few special cases.

1. Continuous saturation of the transition $1 \to 2$. We have:

$$W \to \infty, \quad d\Delta n_{12}/dt = d\Delta n_{34}/dt = 0$$

whence

$$\Delta n_{12} = 0, \quad (\Delta n_{34})_{\text{cont}} = \Delta n_{34}^0\left[1+\frac{T_1}{T_{12}}\frac{1}{1+\beta m}\right]^{-1}. \tag{4}$$

2. Saturation by a short pulse ($\tau \ll T_{12}$). Let $\Delta n_{12} = 0$ and $\Delta n_{34} = \Delta n_{34}^0$ at $t = 0$. Then the restoration of thermal equilibrium after turning off the saturating pulse is described by Eqs. (3) and the amplitudes of the exponentials have a ratio

$$B/A = \beta m. \tag{5}$$

3. Saturation by a long pulse ($\tau \gg T_{12}$). Let $\Delta n_{12} = 0$ and $\Delta n_{34} = (\Delta n_{34})_{\text{cont}}$. The re-establishment of thermal equilibrium after the effect of the long saturating pulse will follow a relaxation curve described by Eqs. (3) with an amplitude ratio:

$$B/A = \beta m T_{12}/(T_1 + T_{12}). \tag{6}$$

The following conclusions can be drawn from the preceding analysis. If one carries out a pulse saturation of the transition $1 \to 2$ and observes the restoration of the paramagnetic resonance line corresponding to this transition following the pulse, then the observed relaxation curve will be described by the sum of two exponentials, one describing the spin–lattice relaxation, and the other, much faster one the spin–spin cross-relaxation. If saturation is by a short pulse, the importance of the cross-relaxation exponential grows with an increase in the multiplicity m of the cross-relaxation transitions between the level pairs E_1, E_2 and E_3, E_4. For saturation with a long pulse the importance of the cross-relaxation exponential is reduced by a factor $(T_1 + T_2)/T_{12}$ compared with the case of a short pulse [cf. Eqs. (5) and (6)]. Note that in Eqs. (5) and (6) the parameter

$$\beta = \Delta n_{34}^0/\Delta n_{12}^0 = \exp\{h(\nu_{43} - \nu_{21})/kT\}$$

depends on temperature; $\beta \approx m$ when $h\nu_{43} \ll kT$.

4. Experimental Results and Discussion

The relaxation processes in chromium corundum were investigated in different lines at crystal trigonal axis orientations relative to the external magnetic field, θ, varying from 0 to 20°. It was found that the shape of the relaxation curves of the restoration of the intensity of the paramagnetic resonance line after the saturating pulse depended strongly on the Cr^{3+} ion concentration. Thus, for saturation by a long pulse the relaxation curves of the various transitions, for samples having concentrations 0.05, 0.4, and 0.65%, were very nearly uni-exponential, whereas for samples having concentrations 0.1 and 0.15% the relaxation curves were described in the majority of the observed transitions by a sum of two exponentials with greatly different time constants. The amplitude ratio of these exponentials depended strongly on the angle θ.

Figure 2 shows oscillograms of the relaxation curves for the transition $-\frac{1}{2} \rightarrow \frac{1}{2}$ at various values of θ, obtained on a sample with Cr^{3+} concentration 0.15% at $T = 4.2°K$. The "fast" exponentials in the relaxation curves were ascribed to the influence of spin–spin cross-relaxation.

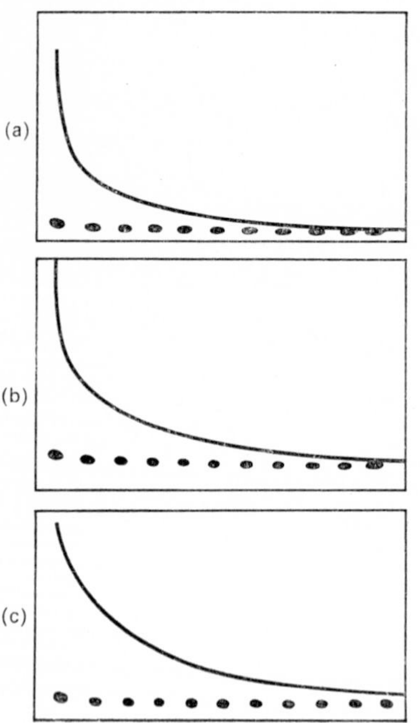

FIG. 2. Oscillograms of relaxation curves of the transition $-\frac{1}{2} \rightarrow \frac{1}{2}$ for a sample having a $Cr_3{}^+$ concentration of 0.15% for different orientations of the crystal: (*a*) $\theta = 0°$, (*b*) $\theta = 10°$, (*c*) $\theta = 15°$. Duration of pulse $\tau = 220$ millisec. The time markers on the baseline are 20 millisec apart.

To investigate the aforementioned effects in greater detail, experiments using different pulse lengths were tried. For short saturating pulses of length $\tau = 0.8$ millisec the importance of the "fast" exponentials increases markedly, confirming that these do characterize the cross-relaxation interaction.

In order to illustrate the influence of saturating pulse length we show in Fig. 3 relaxation curves for the transition $\frac{3}{2} \to \frac{1}{2}$ at $\theta = 0°$

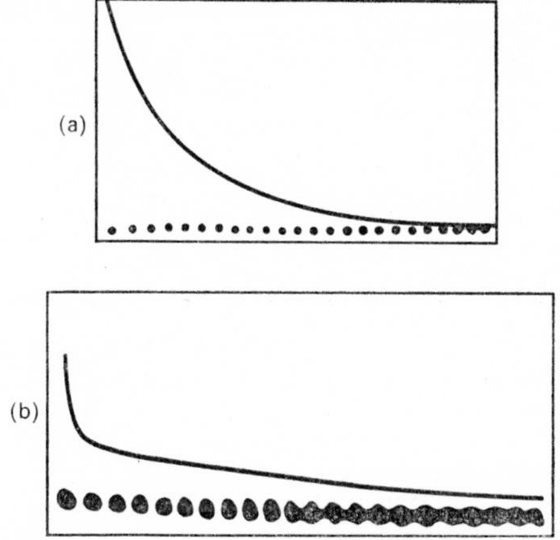

FIG. 3. Relaxation curves of the transition $\frac{3}{2} \to \frac{1}{2}$ ($\theta = 0°$) for different durations of the saturating pulses: (*a*) $\tau =$ 460 millisec, $T = 4.2°$K, (*b*) $\tau = 0.8$ millisec, $T = 1.7°$K. Time markers—20 millisec.

for a 0.15% sample, and for pulse lengths of 460 millisec (Fig. 3*a*) and 0.8 millisec (Fig. 3*b*). In Fig. 3*b* it can be seen that the relaxation curve contains two readily separable exponentials representing fast cross-relaxation and slow spin–lattice relaxation. Similar relaxation curves are also observed for the other transitions in the spectrum.

Tables I and II give the experimental data for the spin–lattice relaxation time T_1 and the spin–spin cross relaxation time T_{12}, as well as the weight of the cross-relaxation exponential $B/(A+B)$ for a 0.15% sample saturated by long and short pulses.

TABLE I. RELAXATION TIMES AND WEIGHT OF THE CROSS-RELAXATION EXPONENTIAL FOR A LONG SATURATING PULSE ($\tau \gg T_{12}$) Cr^{3+} CONCENTRATION 0.15%, $T = 4.2$°K

Transition	θ, deg	T_1, millisec	T_{12}, millisec	$B/(A+B)$, %
$-\frac{1}{2} \to \frac{1}{2}$	0	56.6	4.3	60
$-\frac{1}{2} \to \frac{1}{2}$	10	56	3.7	50
$-\frac{1}{2} \to \frac{1}{2}$	15	56	2.1	20
$\frac{3}{2} \to \frac{1}{2}$	0	96	—	10
$\frac{3}{2} \to \frac{1}{2}$	15	84	—	5
$\frac{1}{2} \to \frac{3}{2}$	0	46	7.8	30
$\frac{1}{2} \to \frac{3}{2}$	10	42	7.1	60
$\frac{1}{2} \to \frac{3}{2}$	15	35	6.1	77
$-\frac{1}{2} \to \frac{3}{2}$	15	40	4.0	46

TABLE II. RELAXATION TIMES AND WEIGHT OF THE CROSS-RELAXATION EXPONENTIAL FOR A SHORT SATURATING PULSE ($\tau \ll T_{12}$) Cr^{3+} CONCENTRATION 0.15%

T, °K	Transition	θ, deg	T_1, millisec	T_{12}, millisec	$B/(A+B)$, %
	$-\frac{1}{2} \to \frac{1}{2}$	0	56.6	2.8	93
	$-\frac{1}{2} \to \frac{1}{2}$	10	56	2	95
4.2	$-\frac{1}{2} \to \frac{1}{2}$	15	56	1	90
	$\frac{1}{2} \to \frac{3}{2}$	0	46	10	63
	$\frac{1}{2} \to \frac{3}{2}$	10	42	1.5	86
	$\frac{1}{2} \to \frac{3}{2}$	15	35	5.5	90
	$-\frac{1}{2} \to \frac{1}{2}$	0	140	3.3	95
	$-\frac{1}{2} \to \frac{1}{2}$	10	140	2.8	95
	$-\frac{1}{2} \to \frac{1}{2}$	15	146	1.6	88
1.7	$-\frac{1}{2} \to \frac{1}{2}$	20	160	1.5	88
	$\frac{3}{2} \to \frac{1}{2}$	0	344	5	69
	$\frac{3}{2} \to \frac{1}{2}$	10	308	4	72
	$\frac{3}{2} \to \frac{1}{2}$	20	316	4	87

TABLE III

θ, deg	Observed transition	Cross-relaxation transition	Ratio of the frequencies of cross-relaxation and observed transitions, m	Calculated weights of the cross-relaxation exponentials (in %) for saturation by	
				long pulses	short pulses
0	$-\frac{1}{2} \to \frac{1}{2}$	$-\frac{3}{2} \to \frac{3}{2}$	3	45	$92(T = 8.2°K)$ $94(T = 2.7°K)$
10	$\frac{1}{2} \to \frac{3}{2}$	$-\frac{1}{2} \to \frac{1}{2}$	2	40	$82(T = 4.2°K)$
15	$-\frac{1}{2} \to \frac{1}{2}$	$-\frac{3}{2} \to -\frac{1}{2}$	2	15	$82(T = 4.2°K)$ $86(T = 1.7°K)$
15	$\frac{1}{2} \to \frac{3}{2}$	$-\frac{3}{2} \to -\frac{1}{2}$	3	63	$92(T = 4.2°K)$

TABLE IV. SPIN-LATTICE RELAXATION TIMES T_1 FOR SAMPLES OF CORUNDUM OF DIFFERENT Cr^{3+} CONCENTRATION

Concentration, %	Transition	θ, deg	T_1, millisec		Concentration, %	Transition	θ, deg	T_1, millisec	
			$T =$ 4.2°K	$T =$ 1.7°K				$T =$ 4.2°K	$T =$ 1.7°K
0.05	$-\frac{1}{2} \to \frac{1}{2}$	0	98	200	0.1	$\frac{3}{2} \to \frac{1}{2}$	0	100	290
	$-\frac{1}{2} \to \frac{1}{2}$	12	—	350		$-\frac{1}{2} \to \frac{3}{2}$	10	59	—
	$\frac{3}{2} \to \frac{1}{2}$	0	208	430		$-\frac{1}{2} \to \frac{3}{2}$	15	43	130
	$\frac{3}{2} \to \frac{1}{2}$	12	130	—					
0.1	$-\frac{1}{2} \to \frac{1}{2}$	0	64	160	0.4	$-\frac{1}{2} \to \frac{1}{2}$	0	20	—
	$-\frac{1}{2} \to \frac{1}{2}$	10	63	145		$\frac{3}{2} \to \frac{1}{2}$	15	21	—
	$-\frac{1}{2} \to \frac{1}{2}$	15	59	145	0.65	$-\frac{1}{2} \to \frac{1}{2}$	0	1	—

In Fig. 4 are shown the energy levels of Cr^{3+} in Al_2O_3 as a function of the intensity of the external magnetic field for various values of the angle θ between this field and the crystal axis. Also shown in these figures are the transitions between which cross-relaxation takes place. As can be seen from Tables I and II, cross-relaxation is most effective for transitions having a whole-number ratio of the splittings between corresponding levels.

However, cross-relaxation also takes place in the case of splittings that are not exact multiples. This leads to a superposition of the effects of cross-relaxation from several transitions. As an example, in the interval $\theta = 0$ to $15°$, cross-relaxation effects between the transitions $-\frac{1}{2} \to \frac{1}{2}$, $-\frac{3}{2} \to \frac{3}{2}$ and $-\frac{3}{2} \to -\frac{1}{2}$ are superposed. This circumstance can probably be explained by the fact that the values of T_{12} as determined with long and short saturating pulses differ from one another for the same observed transition by a factor of 1.5 to 2 (cf. Tables I and II).

The values presented in Tables I and II can be compared with the theory of cross-relaxation considered in Sec. 3. Using the experimentally determined T_1 and T_{12} one can calculate the weights of the cross-relaxation exponential from Eqs. (5) and (6). The results of such calculations for some of the transitions are presented in Table III. A comparison of the data in Tables I, II, and III shows that there is a rather good agreement between the experimental and theoretical values of the amplitudes of the cross-relaxation exponentials, particularly for the cases involving short saturating pulses.

It is interesting to note that the effects of cross-relaxation are strongly pronounced for the $\frac{3}{2} \to \frac{1}{2}$ transition in weak fields. For the transition at $\theta = 20°$ the weight of the cross-relaxation exponential amounts to 87% in saturation by short pulses (see Table II), which agrees with the calculated value if one assumes $m = 2$. However, it can be seen in Fig. 4*d* that there are no cross-relaxation transitions of multiplicity $m = 2$ for this transition. This suggests that besides the cross-relaxation interactions, between levels having a multiple ratio of splittings, considered in Sec. 3, a "combination" type of cross-relaxation interaction is also possible.

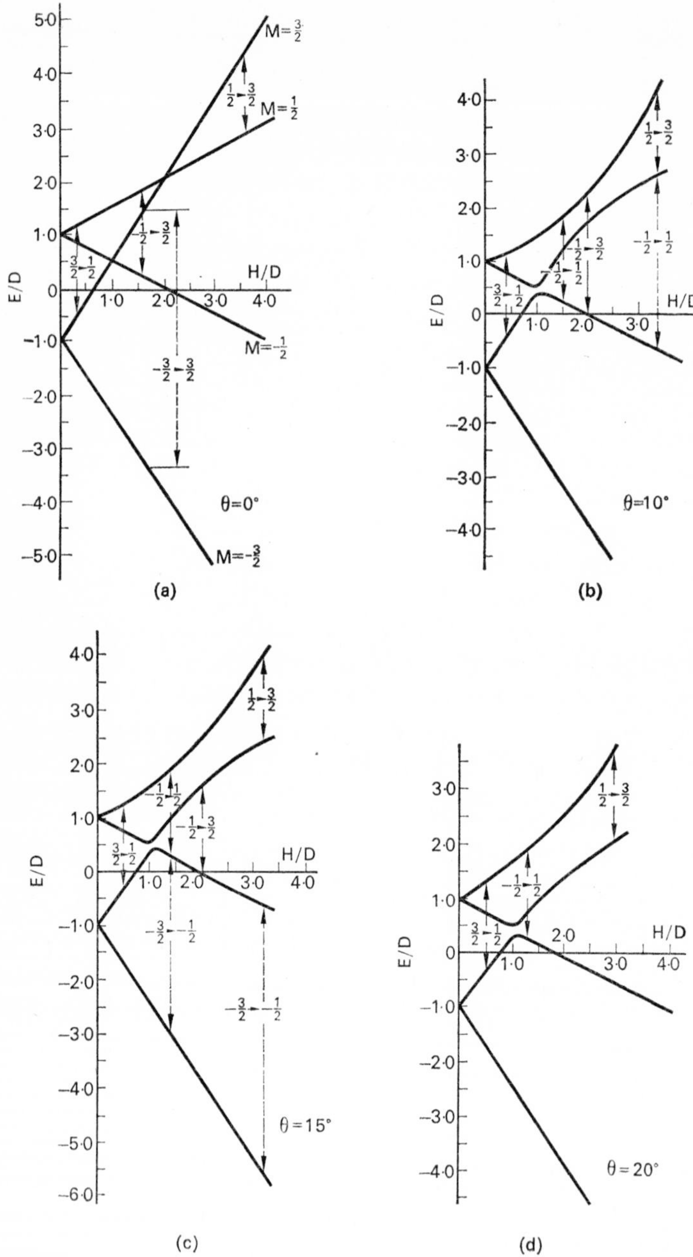
E/D
H/D
M = 3/2
M = 1/2
M = −1/2
M = −3/2
θ = 0°
(a)
θ = 10°
(b)
θ = 15°
(c)
θ = 20°
(d)

Transitions that can occur between levels by such "combination" spin–spin interactions are illustrated in Fig. 5. If the splittings of levels 1, 2, 3, and 4 are such that $2(E_4-E_2) = 2(E_2-E_1)+(E_4-E_3)$, then cross-relaxation can occur as a result of a transition of two ions from level 4 to level 2 accompanied by the transition of two ions from level 1 to level 2 and one ion from level 3 to level

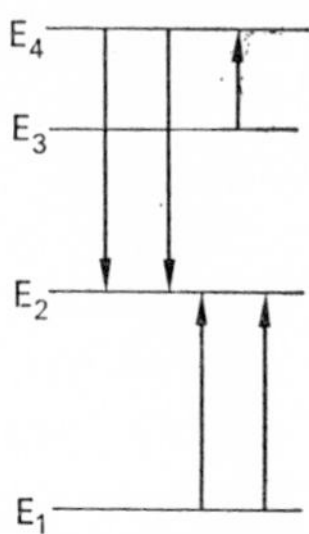

FIG. 5. Transitions for cross-relaxation processes of the "combination" type.

4. The total energy of the spin system is conserved in such transitions. If such "combination" processes are effective, then they can explain the cross-relaxation for the $\frac{3}{2} \rightarrow \frac{1}{2}$ transition. For this transition at $\theta = 20°$ we have the following relation between the splittings of the levels: $2(E_{1/2}-E_{3/2}) = 2(E_{-3/2}-E_{3/2})+(E_{1/2}-E_{-1/2})$.

We shall now present the results of the investigation of relaxation in samples of corundum having Cr^{3+} concentrations of 0.05, 0.4, and 0.65%. For these samples, as we have already noted above, the observed relaxation curves are very nearly uni-exponential and correspond to spin–lattice relaxation. The relaxation times determined by long-pulse saturation are presented in Table IV. Also shown are the values of T_1 for a 0.1% sample. It can be seen from

---

FIG. 4 (*facing*). Energy levels of Cr^{3+} in Al_2O_3 as a function of the applied magnetic field intensity H for different values of the angle θ. D is the crystalline field parameter. The arrows indicate transitions: solid—observed at 9400 Mc, dashed—cross-relaxation.

Table IV that the spin–lattice relaxation times at helium temperatures depend strongly on the Cr^{3+} ion concentration. This fact was established earlier by us[(4)] and was later confirmed in papers by Pace *et al.*[(6)] and by Zverev.[(10)] It has been theoretically discussed on the basis of an assumption of the effect of defects in the crystalline lattice.[(2)] It is also possible that the concentration dependence of the spin–lattice relaxation time in chromium corundum is caused by exchange interactions between Cr^{3+} ions.[(10)] The fact that the paramagnetic resonance line width of Cr^{3+} in Al_2O_3 is not explainable by the theory of dipole broadening [(12)] is possibly connected with the presence of exchange interactions. The temperature dependence of T_1 in the interval 4.2–1.7°K follows the law $T_1 \sim T^{-1}$ within the limits of experimental error ($\sim 10-15\%$), and this indicates that the spin–lattice relaxation at helium temperatures occurs as the result of direct one-phonon processes.

We might remark that the discrepancy at high concentrations between the values for T_1 presented here, which were determined by the pulse saturation method, and the values of T_1 we obtained earlier from experiments using continuous saturation,[(4)] is attributable to the influence of cross-relaxation. At low concentrations, where the cross-relaxation interactions are not effective, both methods give close values for T_1.

Translated by L. M. Matarrese.

References

(1) Bloembergen, Shapiro, Pershan and Artman, *Phys. Rev.* **114,** 445 (1959).
(2) B. I. Kochelaev, *DAN SSSR* **131,** 1053 (1960); *Soviet Phys.-Doklady* **5,** 349 (1960).
(3) P. W. Anderson, *Phys. Rev.* **114,** 1002 (1959).
(4) A. A. Manenkov and A. M. Prokhorov, *JETP* **38,** 729 (1960); *Soviet Phys. JETP* **11,** 527 (1960).
(5) Pace, Sampson and Thorp, *Phys. Rev. Letters* **4,** 18 (1960).
(6) Pace, Sampson and Thorp, *Proc. Phys. Soc. (London)* **77,** 257 (1961).
(7) B. Bolger and B. J. Robinson, *Physica* **26,** 133 (1960).
(8) R. A. Armstrong and A. Szabo, *Can. J. Phys.* **38,** 1304 (1960).

(9) J. E. GEUSIC, *Phys. Rev.* **118,** 129 (1960).
(10) G. M. ZVEREV, *JETP* **40,** 1667 (1961); *Soviet Phys. JETP* **13,** 1175 (1961).
(11) A. A. MANENKOV and V. A. MILYAEV, *JETP* **41,** 100 (1961); *Soviet Phys. JETP* **14,** 75 (1962).
(12) A. A. MANENKOV and V. B. FEDOROV, *JETP* **38,** 1042 (1960); *Soviet Phys. JETP* **11,** 751 (1960).

PAPER 10*

Microwave Generation in Ruby due to Population Inversion produced by Optical Absorption†

D. P. DEVOR, I. J. D'HAENENS and C. K. ASAWA

Hughes Research Laboratories, Malibu, California
(Received April 30, 1962)

Microwave amplification and generation by the stimulated emission of radiation (maser) were observed in ruby as a result of population inversion produced in the ground state of Cr^{3+} by the absorption of the coherent optical emission from a second ruby (optical maser). The maser crystal was oriented in a magnetic field of about 6700 Oe to obtain a transition between a ground $^4A_2$ Zeeman sublevel and an excited $\bar{E}(^2E)$ Zeeman sublevel which would match a spectral component of the output of the optical maser. The arrangement of the experimental apparatus is shown schematically in Fig. 1.

Detection of the interaction of optical and paramagnetic transitions in the sharp line spectra of ions in crystalline solids has been accomplished previously through the observation of the effect of the paramagnetism on the optical signals.[(1-3)] Wieder[(1)] attempted to directly detect the effect on paramagnetism of depopulation in the $^4A_2$ state in ruby due to the absorption of the R-line fluorescent light. Since the optical spectrum of the ruby optical maser results from R-line emission, the optical maser offered a considerably

* *Phys. Rev. Letters* **8**, 432–5 (1962).

† Work supported by the U.S. Army Signal Corps under Contract DA 36–039 SC–87221.

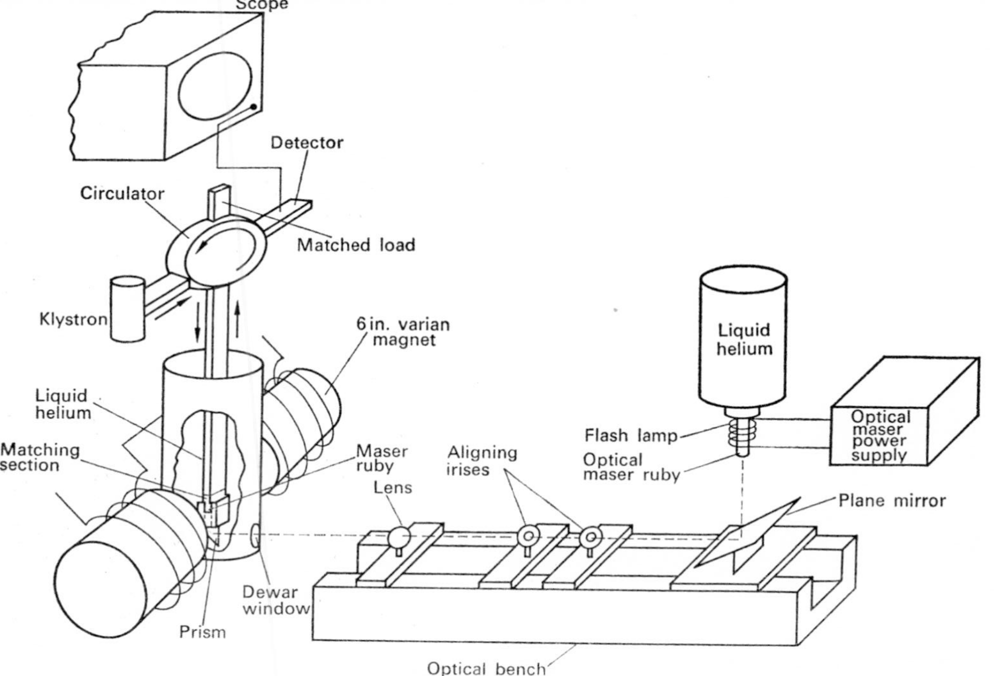

FIG. 1. Arrangement of experimental apparatus.

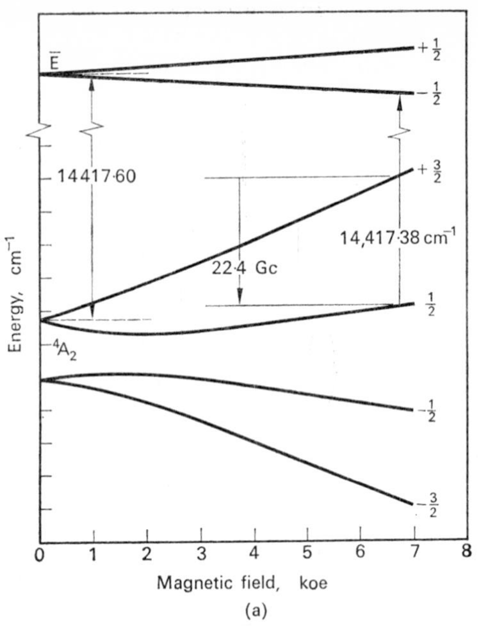

(a)

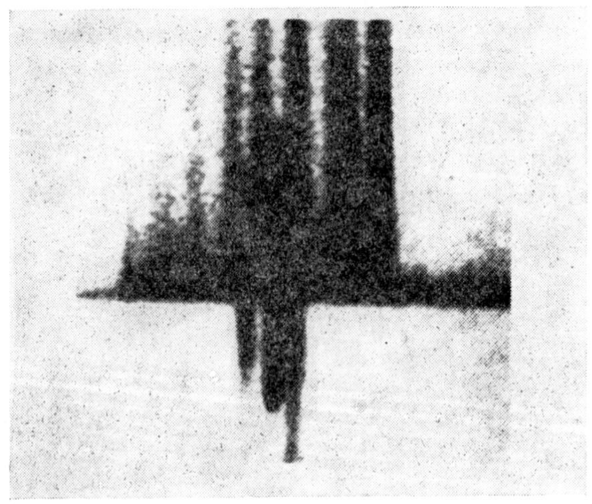

Wavelength λ Increasing
(b)

better optical source (in terms of collimation, spectral purity, and power output) for such an experiment, not only to detect perturbation of the ground-state populations, but also to produce population inversion.

D'Haenens and Asawa[4] have observed stimulated optical emission in ruby to shift to longer wavelengths from the fluorescent emission. In the present experiment, the optical-maser crystal, initially cooled by conduction to liquid-helium temperature, produced a pump signal of 6934.188 Å (air wavelength) from the $\bar{E}(^2E) \rightarrow \pm\frac{1}{2}(^4A_2)$ transition (spin states correspond to high-field assignments). The optical-maser output was unpolarized.

The separation of the $\bar{E}(^2E)$ and the $\pm\frac{1}{2}(^4A_2)$ states in the maser crystal was taken as the fluorescent wavelength of 6934.082 Å at 4.2°K. The g value of the $\bar{E}$ Kramers' doublet was obtained from the work of Geschwind, Collins, and Schawlow,[2] and the Zeeman structure of the $^4A_2$ ground state was known from the analysis of Chang and Siegman.[5] From their data we calculated that the magnetic field at an angle of 67 degrees, with respect to the crystalline c axis, split the $\bar{E}$ and $^4A_2$ states as shown in Fig. 2(*a*). At 6700 Oe, the $+\frac{1}{2}(\bar{E}) \rightarrow +\frac{1}{2}(^4A_2)$ transition matched the $\bar{E} \rightarrow \pm\frac{1}{2}(^4A_2)$ component of the optical-maser spectrum, since the latter transition was 0.22 cm^{-1} less than the fluorescent wave number in zero field. The $+\frac{3}{2}(^4A_2) \rightarrow +\frac{1}{2}(^4A_2)$ microwave transition occurred at 22.4 Gc/sec. Figure 2(*b*) shows a comparison of

---

FIG. 2 (*facing*). Comparison of the spectra of ruby in fluorescence and in stimulated emission under conditions suitable for optical pumping. (*a*) Zeeman structure of Cr^{3+} ruby. The crystalline c axis is oriented at 67 degrees with respect to the magnetic field. The g factor of the $\bar{E}$ state is 0.956, as given by Geschwind *et al.*[2] (*b*) Comparison of spectrographs of ruby in fluorescence (upper eight lines) and in stimulated emission at a temperature of 4.2°K. The fluorescence was observed with the c axis at 70 degrees with respect to a magnetic field of 6500 Oe. The short line is the neon reference in ninth order corresponding to $\lambda = 6934.0831$ Å. Spectra were taken with a special Harrison grating.

the fluorescence and optical-maser emission spectra under approximately such conditions. The spectral agreement of the $-\frac{1}{2}(\bar{E}) \rightarrow +\frac{1}{2}(^4A_2)$ transition in fluorescence with the lower energy optical-maser line indicated that the magnitude and orientation of the magnetic field were not particularly critical.

The microwave resonant structure consisted of a section of 0.050- by 0.130-inch waveguide which was beyond cutoff at 22.4 Gc/sec and was partially loaded with a 0.078-inch length of ruby. The structure acted as a ruby-loaded microwave reflection cavity resonant at 22.4 Gc/sec in a perturbed TE_{011} mode. At liquid-nitrogen temperature, the quality factor Q_0 of this structure, exclusive of magnetic losses and external circuit losses, was about 1600. The filling factor was calculated to be 96%.

The conditions for power gain and/or oscillation in a solid-state maser are given by Bloembergen[(6)] in terms of the magnetic quality factor Q_m. To overcome the microwave circuit losses, the quality factor Q_0 of the structure must exceed $|Q_m|$ when population inversion is obtained. The population of the $\bar{E}$ state is extremely small at 4.2°K and, consequently, saturation of the $+\frac{1}{2}(^4A_2) \rightarrow -\frac{1}{2}(\bar{E})$ pump transition depletes the $+\frac{1}{2}(^4A_2)$ state of one-half the Boltzmann population. The maser crystal was cut from 0.05 wt.% $Cr_2O_3 : Al_2O_3$ ruby, giving a Boltzmann population of 3.06×10^{16} in the $+\frac{1}{2}(^4A_2)$ level. Thus, the absorption of 1.53×10^{16} photons, or 4.39×10^{-3} joule of pump signal, is required for saturation of the $\frac{1}{2}(^4A_2) \rightarrow -\frac{1}{2}(\bar{E})$ transition; this amount of energy is orders of magnitude less than that available from the optical maser. Using appropriate data for ruby and the present microwave cavity, the magnetic Q obtained from the saturation of the $\frac{1}{2}(^4A_2) \rightarrow \frac{1}{2}(\bar{E})$ transition is $-Q_m \simeq 75$, which is considerably less than the Q_0 of the microwave resonant structure.

The experimental results are shown in Fig. 3. The klystron signal generator was frequency modulated at a repetition rate which allowed us to sample the relative populations of the $^4A_2$ sub-levels while the microwave system was being optically pumped. This low effective duty cycle was chosen to avoid microwave saturation effects. The cavity was undercoupled in the absence of

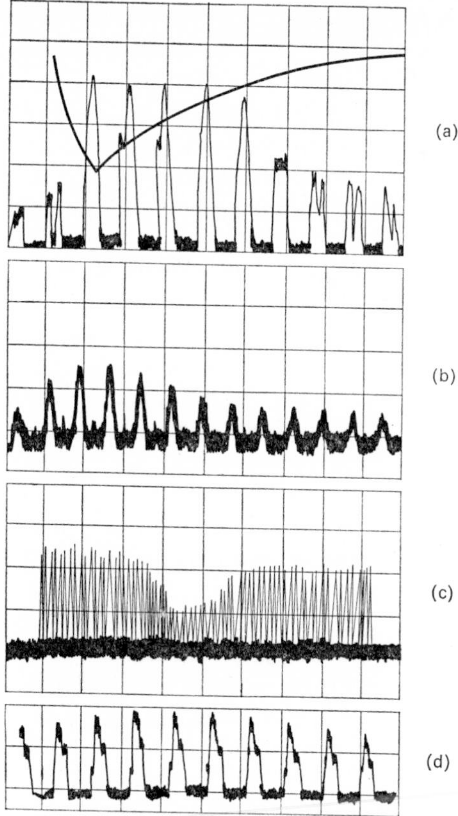

FIG. 3. Experimental results. (*a*) Optical-maser output and a 200-μsec-delayed signal of reflected microwave power from the microwave cavity showing amplification at 22.4 Gc/sec. The optical-maser output was about 2.5 joules of which 1 joule was estimated to be incident on the maser crystal. Time scale = 200 μsec/division; input microwave power level $\approx$ 50 μW. (*b*) Simultaneous maser oscillation and amplification. Conditions as in (*a*) except microwave coupling to the maser crystal was reduced; also, the optical-maser output was increased to about 4 joules of which about 2.8 joules were estimated to be incident on the maser crystal. (*c*) Maser oscillation observed with no input microwave signal. Time scale = 50 μsec/division. (*d*) Increased microwave absorption at the $-\frac{1}{2}(^4A_2) \rightarrow +\frac{1}{2}(^4A_2)$ transition due to depopulation of the $+\frac{1}{2}(^4A_2)$ level. All conditions were as in (*a*).

paramagnetic resonance. The traces in Fig. 3(*a*) are recordings of the optical-maser output and a 200-μsec-delayed display of the microwave reflected power. The first klystron sweep shows the ruby sample absorbing at the $+\frac{1}{2}(^4A_2) \rightarrow +\frac{3}{2}(^4A_2)$ transition; the second sweep shows the reduction in magnetic losses as $1/Q_m$ goes to zero (or becomes slightly negative) due to depopulation of the $+\frac{1}{2}(^4A_2)$ level; on the third sweep, $1/Q_m$ is sufficiently negative to produce amplification.

In Fig. 3(*b*) the microwave coupling to the maser crystal was reduced, the optical-maser output was increased, and the klystron power level was the same as that in Fig. 3(*a*). Figure 3(*b*) shows simultaneous amplification and oscillation. For the trace in Fig. 3(*c*) the klystron was turned off and oscillation was produced. The pulsed nature of the oscillation was the same as that observed in microwave-pumped ruby masers by Makhov *et al.*[(7)] and Foner *et al.*[(8)] The $-\frac{1}{2}(^4A_2) \rightarrow +\frac{1}{2}(^4A_2)$ resonance was observed with a small increase in magnetic field. A recording of this resonance, when the optical maser illuminated the maser crystal, showed that the microwave absorption increased markedly [Fig. 3(*d*)] and thus confirmed the pumping of the $+\frac{1}{2}(^4A_2)$ level.

The microwave structure appeared to have two resonant modes separated by 30 or 40 Mc/sec, possibly as a result of the anisotropy of the dielectric constant of sapphire and a slight misalignment of the crystalline *c* axis from the rectangular axis of the cut crystal. The maser was occasionally observed to amplify in both modes simultaneously, although not with equal gain.

With the technique described here, amplification and oscillation at frequencies considerably higher than 22.4 Gc/sec are possible; thus, a brief examination of the energy levels of ruby indicates that amplification at 57 Gc/sec in a magnetic field of 21 000 Oe could be obtained. In addition, the excited state can be selectively populated to allow direct observation of paramagnetic resonance in this state. Furthermore, maser gain was observed for some time after the loss of the liquid helium in the maser Dewar, thus indicating the possibility of operating at temperatures higher than 4.2°K.

The efficacy of the optical maser in an optical-pumping experiment was, to our knowledge, first considered by T. H. Maiman and R. H. Hoskins. Considerable help with the microwave resonant structure was obtained from J. E. Keifer and F. E. Goodwin.

References

(1) Irwin Weider, *Phys. Rev. Letters* **3**, 468 (1959).
(2) S. Geschwind, R. J. Collins and A. L. Schawlow, *Phys. Rev. Letters* **3**, 545 (1959).
(3) J. Brossel, S. Geschwind and A. L. Schawlow, *Phys. Rev. Letters* **3**, 548 (1959).
(4) I. J. D'Haenens and C. K. Asawa (to be published).
(5) W. S. Chang and A. E. Siegman, Stanford Electronic Laboratory, Technical Report No. 156–2, Stanford University, California (unpublished); also as reproduced by J. Weber, *Revs. Modern Phys.* **31**, 681 (1959).
(6) N. Bloembergen, *Phys. Rev.* **104**, 324 (1956).
(7) G. Makhov, C. Kikuchi, J. Lamb and R. W. Terhune, *Phys. Rev.* **109**, 1399 (1958).
(8) S. Foner, L. R. Momo and A. Mayer, *Phys. Rev. Letters* **3**, 36 (1959).

Comments on Papers 11–12

THESE papers describe the first successful demonstration of a maser operating on the Bloembergen principle. In many ways this was a remarkable experiment and that it succeeded is a great tribute to the skill of these workers. The material chosen was lanthanum ethyl sulphate containing 0·5% gadolinium as the active ion (i.e. Gd atoms; La atoms = 1 : 200). The ground state of Gd^{+++} is $^8S_{7/2}$ and thus has an eightfold spin degeneracy, the spacing of the eight levels being rather similar. As already mentioned (Section 2.6) it was possible to obtain inversion in this system only by artificially increasing the relaxation rate w_{23} by exploiting a cross-relaxation process between the gadolinium ion and a second ionic species, cerium, which was introduced into the lattice for this explicit purpose.

The presentation of the results of the experiment is unusual and calls for comment. The power reflected from the cavity at a frequency of 9 GHz (kMc/s) is displayed as the magnetic field is swept (presumably slowly) to cover three resonance lines $-\frac{7}{2} \rightarrow -\frac{5}{2}$; $-\frac{5}{2} \rightarrow -\frac{3}{2}$ and $-\frac{3}{2} \rightarrow -\frac{1}{2}$. The effects of the simultaneously applied pump frequency are apparent on the $-\frac{7}{2} \rightarrow -\frac{5}{2}$ and the $-\frac{5}{2} \rightarrow -\frac{3}{2}$ lines—the latter by design, the former by chance. The effects of the simultaneous pump and signal transitions are observed over a range of fields which is very much less than the apparent width of the signal transition; to some extent this would be expected since the width (in oersteds) of the pump transition is likely to be less than that of the signal. Nevertheless, the very narrow range of fields over which simultaneous interaction is observed leads one to suppose that there must be some inhomogeneous broadening of either the pump or signal transition or both.* In Fig. 1 the authors give an indication of the splittings of the ground state levels at their desired operating point together with measured values of the spin–lattice relaxation times between the various levels. The effect of cross relaxation in speeding up the relaxation process between the $-\frac{1}{2}$ and $-\frac{3}{2}$ levels is clearly indicated here but one should not suppose that the measured values of T_1 between the various levels necessarily give an indication of the individual direct transition probabilities (the w_{ij}) due to spin–lattice interaction between these levels. The observed relaxation in a multilevel system is due to a combination of all the possible individual relaxation processes.

The authors did not measure the gain of their maser but observed conditions under which stable gain was possible and they also noted the onset of oscillation at the signal frequency.

* Further discussion of this point with Dr. H. E. D. Scovil has confirmed that there was a substantial inhomogeneous contribution to the line broadening in the B.T.L. experiment.

PAPER 11*

Electron Spin Relaxation Times in Gadolinium Ethyl Sulfate

G. FEHER and H. E. D. SCOVIL

Bell Telephone Laboratories, Murray Hill, New Jersey

(Received December 3, 1956)

THE recent proposal by Bloembergen[1] to utilize a paramagnetic salt to amplify or produce microwave power led us to investigate the electron spin relaxation times in dilute gadolinium ethyl sulfate having cerium as an additional impurity. The preliminary results of these investigations are being reported in this letter. The successful operation of a maser incorporating this material is described in the accompanying letter.[2]

The trivalent gadolinium ion is in an ^8S ground state having seven electrons in a half-filled $4f$ shell. The energy levels were investigated in detail by Bleaney, Scovil and Trenam.[3] The essential features are that the line is split by a fine structure term into seven lines whose spacing varies approximately as $(3\cos^2\theta-1)$ where θ is the angle between the dc magnetic field H_0 and the crystalline field axis. Figure 1 shows the position of four of the lines, the remaining three being omitted for clarity. The inequality of the spacings and the fact that the lines do not cross at 55° can be taken into account by higher order corrections. The paramagnetic

* *Phys. Rev.* **105,** 760–2 (1957).

(1) N. Bloembergen, *Phys. Rev.* **104,** 324 (1956).

(2) Scovil, Feher and Seidel, *Phys. Rev.* **105,** 762 (1957).

(3) Bleaney, Scovil and Trenam, *Proc. Roy. Soc.* (*London*) **A223,** 15 (1954).

resonance of cerium was first investigated by Bogle, Cooke and Whitley.[4] Its ground state transition is also indicated in Fig. 1.

The crystals investigated were magnetically diluted with isomorphous lanthanum ethyl sulfate in the ratio Gd : La = 1 : 200

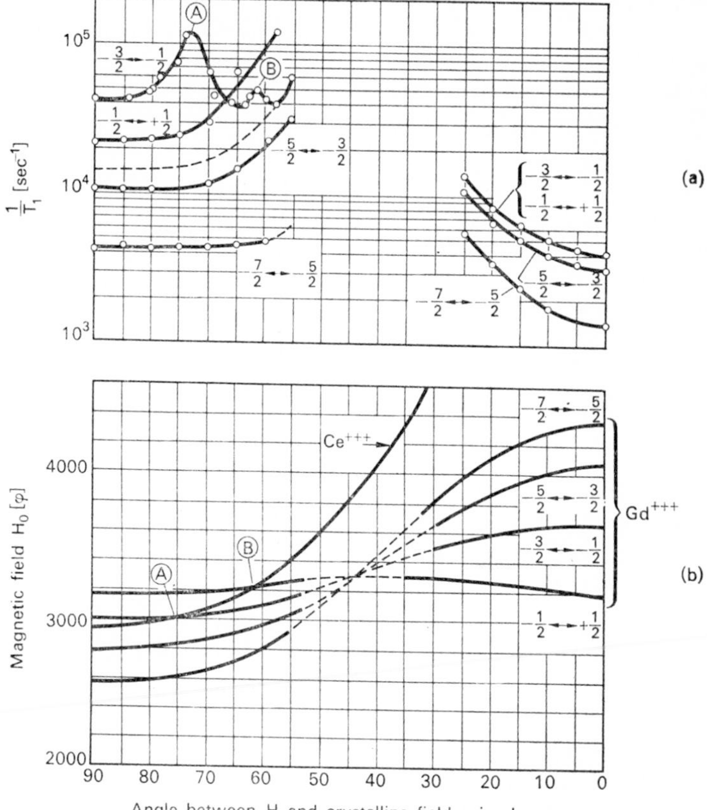

FIG. 1. (*a*) Inverse relaxation times for the different Gd transitions *vs* angle between H_0 and crystalline axis. Note the reduction in relaxation time when the Gd and Ce transitions overlap. (*b*) Magnetic field H_0 at which Gd and Ce transitions occur *vs* angle between H_0 and crystalline axis.

[4] Bogle, Cooke and Whitley, *Proc. Roy. Soc.* (*London*) **A64,** 931 (1951).

and Ce : La = 1 : 500. The size of the crystal used was approximately 1 mm × 1 mm × 0.5 mm. The small size was chosen to avoid perturbations on the cavity.

The relaxation times were investigated by the power saturation method.[5] This procedure gives a unique relaxation time for a system which has only two energy levels (i.e., spin $\frac{1}{2}$). In a more complicated system one might not be able to associate a single relaxation time with each pair of levels.[6] However, the microwave power at which χ'' drops by a given fraction is still a measure of the rate at which energy is carried to the lattice via all relaxation processes. In all our experiments we measured the power at which the saturation parameter $S = \chi''(H_1)/\chi''(H_1 = 0)$ dropped to 0.316 (10 db). In order to calculate the spin–lattice relaxation time T_1 from this measurement, we write for the saturation parameter:[5,7]

$$S = \frac{\chi''(H_1)}{\chi''(H_1 = 0)} = \frac{1}{1+(W_{M\to M-1}/W_{\text{S.L.}})}$$

$$= \frac{1}{1+\frac{1}{4}(\gamma H_1)^2[S(S+1)-M(M-1)]T_1 T_2},$$

where $W_{M\to M-1}$ is the transition probability for $\Delta M = \pm 1$ in an external magnetic field of amplitude H_1, $1/W_{\text{S.L.}}$ is twice the spin–lattice relaxation time T_1, γ is the effective gyromagnetic ratio and $T_2 = 1/\pi\Delta\nu$, $\Delta\nu$ being the width at half maximum power absorption. The experimentally determined T_2 was 8×10^{-9} sec. The line width is due primarily to interactions of the electron spin with the magnetic moments of the neighboring protons, although the electron dipole–dipole interactions are of the same order. It is therefore not surprising that the line is homogeneously broadened as found from its saturation behavior.[8]

The results of the relaxation time measurements at 1.2°K and

(5) Bloembergen, Purcell and Pound, *Phys. Rev.* **73,** 679 (1948).

(6) J. P. Lloyd and G. E. Pake, *Phys. Rev.* **94,** 579 (1954).

(7) A. H. Eschenfelder and R. T. Weidner, *Phys. Rev.* **92,** 869 (1953).

(8) A. M. Portis, *Phys. Rev.* **91,** 1071 (1953).

9000 Mc/sec are shown in Fig. 1. The most striking feature is exhibited by the $-\frac{3}{2} \rightarrow -\frac{1}{2}$ line. When its resonance frequency is equal to that of cerium (see point *A*, Fig. 1), its relaxation time is reduced by about a factor of 7. This assumes that in the absence of cerium the gadolinium relaxation time would follow the dotted line. A reduction of the relaxation time of Gd by Ce has been reported earlier by Bleaney, Elliott and Scovil.[(9)] Their investigations were carried out on a concentrated Ce salt at 14°K. Under those circumstances all Gd levels were affected simultaneously, which is undesirable for the operation of a maser. From those experiments one concludes that the relaxation process proceeds in two steps. First, there is a Gd–Ce spin–spin flip which is energetically most favorable, when the resonance frequencies of the two ions are equal. This process proceeds in a characteristic time essentially given by T_2, which in our case is $\sim 10^{-8}$ sec. In the second step the Ce relaxes via its own spin–lattice relaxation time which at 1.2°K was found to be about three orders of magnitude longer than T_2.

Another reduction in the relaxation time is observed when we approach a point at which the Gd transitions overlap. (see Fig 1). In this case the Gd transitions whose resonant frequencies lie nearest to the saturated line take the place of the Ce. It should be pointed out, however, that an unknown impurity with a broad resonance, could account for a similar behavior.

The third distinct relaxation process occurs at a point where the Ce transition overlaps a Gd transition differing by $\Delta M = 1$ from the saturated line (see point *B*, Fig. 1). This involves a three-step process. First the saturation is partially passed on from the $-\frac{3}{2} \rightarrow -\frac{1}{2}$ levels to the $-\frac{1}{2} \rightarrow +\frac{1}{2}$ levels via a spin–spin interaction. The $-\frac{1}{2} \rightarrow +\frac{1}{2}$ levels then relaxes via the Ce as discussed earlier.

In the absence of strong interactions between different levels ($\theta = 0$ in Fig. 1), the relaxation times are approximately proportional to the inverse transition probabilities as one would expect.

We would like to thank Dr. P. W. Anderson for helpful discussions and Mr. E. A. Gere for his assistance in the experiments.

(9) Bleaney, Elliott and Scovil, *Proc. Phys. Soc. (London)* **A64**, 933 (1951).

PAPER 12*

Operation of a Solid State Maser

H. E. D. Scovil, G. Feher and H. Seidel

Bell Telephone Laboratories, Murray Hill, New Jersey

(Received December 3, 1956)

A MASER of the same type as that proposed by Bloembergen[(1)] has been successfully operated at 9 kMc/sec. Since the basic theory has been covered in the reference, it will not be reviewed here.

We require a magnetically dilute paramagnetic salt having at least three energy levels whose transitions fall in the microwave range and which may be easily saturated. The ion $Gd^{+++}|4f^7, {}^8S\rangle$ seems a suitable choice since its eight energy levels give the choice of several modes of maser operation. Of the three salts of Gd^{+++} which have been investigated by paramagnetic resonance[(2)] the diluted ethyl sulfate appears very desirable. This salt has been investigated in detail by Bleaney *et al.*,[(3)] Buckmaster[(4)] and Feher and Scovil.[(5)]

If an external magnetic field is applied perpendicular to the magnetic axis, the spin Hamiltonian may be written[(3)]

$$\mathscr{H} = g\beta H_0 \cdot S_z - \tfrac{1}{2}B_2{}^0[S_z{}^2 - \tfrac{1}{3}S(S+1)] + \tfrac{1}{4}B_2{}^0[S_+{}^2 + S_-{}^2], \tag{1}$$

* *Phys. Rev.* **15,** 762–3 (1957).

(1) N. Bloembergen, *Phys. Rev.* **104,** 324 (1956).

(2) K. D. Bowers and J. Owen, *Repts. Progr. in Phys.* **18,** 304 (1955).

(3) Bleaney, Scovil and Trenam, *Proc. Roy. Soc.* (*London*) **A223,** 15 (1954).

(4) H. A. Buckmaster, *Can. J. Phys.* **34,** 150 (1956).

(5) G. Feher and H. E. D. Scovil, preceding Letter [*Phys. Rev.* **105,** 760 (1957)].

where some small terms have been neglected, $g = 1.99$, $B_2^0 \approx 0.02\ \text{cm}^{-1}$, and the axis of quantization is parallel to H_0. The first term is the usual Zeeman energy and is varied to bring the transitions to the desired operating frequency. The second term disturbs the equality of the level spacings (essential for the device) as shown in Fig. 1. The third term admixes states, thereby permitting

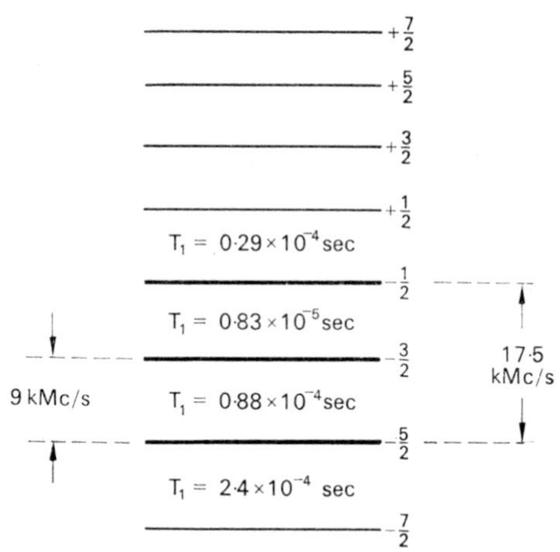

FIG. 1. The energy levels of the ground state of Gd^{+++} in the ethyl sulfate for a large applied magnetic field. The heavy lines identify the maser levels. Spin–lattice relaxation times between levels are shown.

$\Delta S_z = \pm 2$ transitions which are also essential. The angle between the dc magnetic field and the microwave magnetic field should be zero for the $\Delta S_z = \pm 2$ transitions and 90° for the $\Delta S_z = \pm 1$ transitions. A convenient compromise of 45° between both microwave fields and H_0 was chosen for the structure employed.

The negative temperature (a term introduced by Purcell and Pound[(6)] to designate the fact that a higher energy level is more

[(6)] E. M. Purcell and R. V. Pound, *Phys. Rev.* **81,** 279 (1951).

densely populated than the lower one) at complete saturation of the $\Delta S_z = \pm 2$ transition will depend essentially upon two parameters: the separations of the energy levels and the ratio of the relaxation times. In a given material the first parameter is fixed. Our attempts were directed toward varying the second parameter in order to obtain lower negative temperatures. A relaxation time ratio of 1 : 10 between two neighboring transitions was obtained by introducing cerium into the crystal.[5] In order to obtain the full benefit of this large relaxation time ratio for a 9-kMc/sec maser, a dc magnetic field of 2850 oersteds was applied at an angle of 17° from the perpendicular direction of the crystal.[5] Although Eq. (1) refers to the perpendicular direction, the energy levels and transition probabilities are only slightly modified at this small angle. A 90-mg (8% filling factor) lanthanum ethyl sulfate crystal containing $\approx 0.5\%$ Gd^{+++} and $\approx 0.2\%$ Ce^{+++} was used in contact with liquid helium at 1.2°K. A saturating magnetic field at 17.52 kMc/sec was used to induce transitions between the $|-5/2\rangle$ and $|-\frac{1}{2}\rangle$ states as shown in Fig. 1. The signal at 9.06 kMc/sec was applied between the $|-5/2\rangle$ and $|-3/2\rangle$ states. The maser embodies a microwave cavity simultaneously resonant at these two frequencies. The almost critically coupled 9-kMc/sec cavity had a loaded $Q \approx 8000$. The 17.5-kMc/sec cavity perversely supporting a spurious mode provided a $Q \approx 1000$; this fortunately proved sufficient.

Figure 2 shows the 9-kMc/sec monitoring signal reflected from the cavity as a function of H_0. In the first trace three $\Delta S_z = \pm 1$ transitions are shown, the peaks representing essentially complete reflection as a result of the high magnetic losses associated with the material. The observed resonance line appears broadened since the absorption is not a small perturbation on the cavity as resonance is approached. The succeeding traces show the reflections associated with the $|-5/2\rangle \rightarrow |-3/2\rangle$ transition as the 17.5 kMc/sec power is increased. In the third trace the salt is lossless, corresponding to an essentially infinite spin temperature. The fourth trace shows the onset of negative spin temperatures and the partial overcoming of the losses associated with the empty

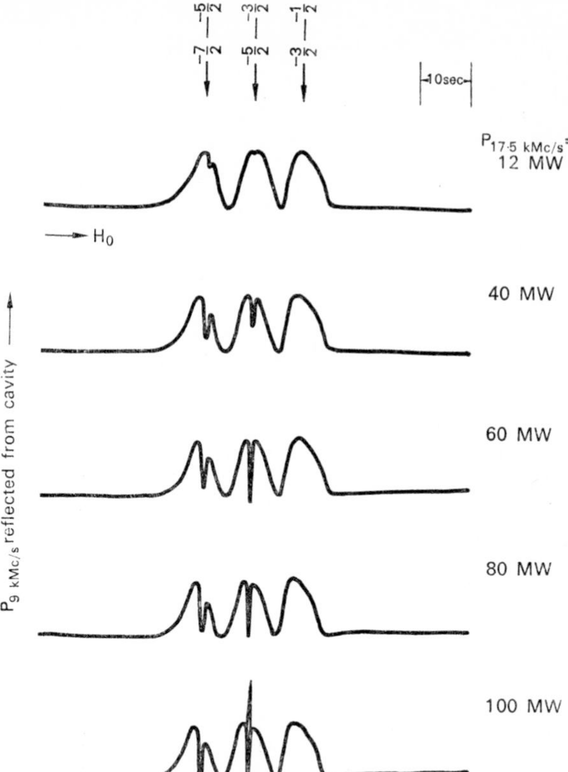

FIG. 2. The power reflected from the 9-kMc/sec cavity as the magnetic field was swept to cover three $\Delta S_z \pm 1$ transitions for different 17.5-kMc/sec power levels. The spacing between two lines is about 200 oersteds.

cavity. In the fifth trace the reflected power exceeds the incident power and oscillations have commenced. Before oscillations commence, a region of amplification must exist. Figure 3 shows the last trace on an expanded time scale.

At this stage, the 9-kMc/sec monitoring signal was turned off. The dc magnetic field was adjusted to a value resulting in maximum

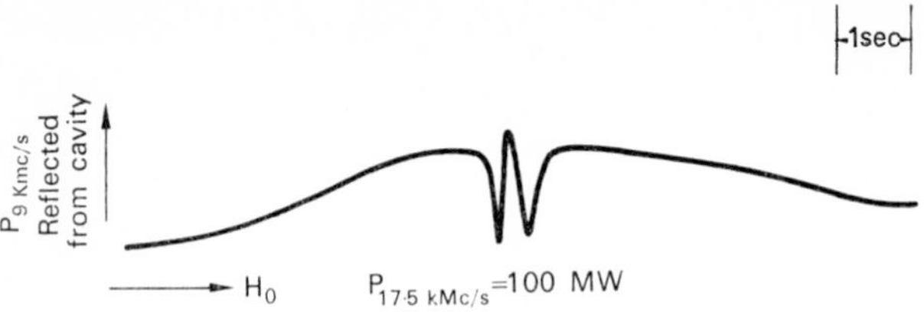

FIG. 3. The central line of the last trace of Fig. 2. is shown on an expanded time scale.

9-kMc/sec output power from the oscillating maser. The power output was measured with a barretter as a function of the saturating 17.5-kMc/sec power. The results are shown in Fig. 4.

The required saturating power could be materially reduced by the use of a 17.5-kMc/sec cavity having a higher Q. The purpose of this work was merely to show the feasibility of this device.

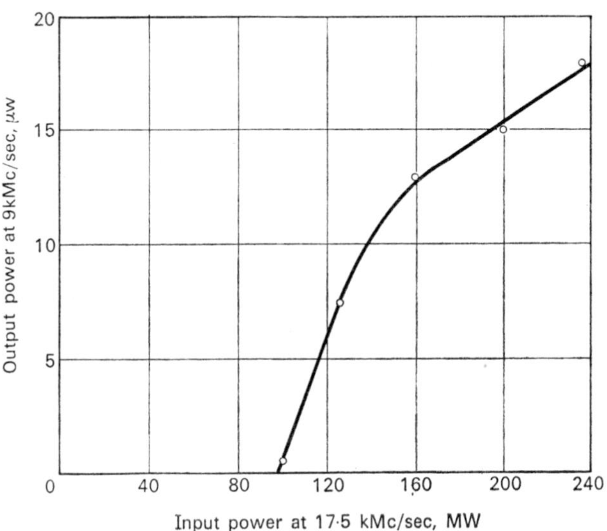

FIG. 4. The 9-kMc/sec output power of the oscillating maser as a function of the saturating power.

We should like to acknowledge the generous cooperation of many members of the Bell Telephone Laboratories, and in particular, to thank Mr. M. G. Gussak.

Comments on Papers 13–14

In these two papers, originating from the Lincoln Laboratory, the authors have set out to do more than just demonstrate the validity of the Bloembergen principle and they describe an actual maser amplifier giving measurements of gain, bandwidth, saturation characteristics and, in the subsequent letter, noise temperature.

The active ion employed is Cr^{3+} in a host lattice of potassium cobalticyanide ($K_3Co(CN)_6$). In this material the chromium ion is in a situation which is not dissimilar to that in ruby except that, whereas in ruby the Cr^{3+} ions are each surrounded by an octahedron of oxygen ions in the cyanide, it is (CN) complexes which lie at the extremities of the octahedron, the symmetry of the crystal field is lower than in ruby and this accounts for the appearance of the extra E terms in the spin Hamiltonian (Chapter 2).

There are two sites per unit cell the magnetic axes of which are slightly differently oriented with respect to the crystallographic axes, a, b, c. Clearly the two sites will appear identical if the applied magnetic field lies in either the *ac*- or *bc*-planes. The normal growth habit of the crystals is with pronounced a faces which tends to make *bc* plane operation easier. ($K_3(Cr)(Co)(CN)_6$) crystals having pronounced b-faces can be prepared and such crystals favour *ac*-plane operation which analysis and experiment subsequent to the work of McWhorter and Meyer has shown to have marked advantages over the *bc*-plane.

The experiments of McWhorter and Meyer went a long way towards demonstrating the validity of the analysis of cavity maser operation but it is important to realise that their apparatus was designed solely to assess the performance of a cavity maser and is not suitable for an actual working maser. In a reflection cavity maser to be used in a system the input and output are separated by means of a low loss circulator which would replace the 30 dB directional coupler employed by McWhorter and Meyer (their Fig. 7). The reader will observe that the sequence of operations illustrated in Fig. 9 of McWhorter and Meyer's paper is precisely similar to the undercoupled cavity behaviour which we discussed earlier.

Too much weight should not be attached to the calculation of gain bandwidth product given in this paper; the expression (8) for P_m cannot be directly evaluated as we do not know the relative magnitude of the w_{ij} and there is no justification for the assumption that the w_{ij} are equal (except perhaps that it gives gain bandwidth products in broad accord with experiment!). The expression given for W_{23} is also an approximation in this case although the results of proper calculation made by Chang and Siegman (1958*a*) indicate that it is not a bad approximation. In their discussion of saturation McWhorter and Meyer suggest that $w_{21} = w_{32} = 1/2T_1$. For a two-level system a relationship of this nature holds between the spin–lattice relaxation time and the thermal transition probability; however, for a four-level system even if we assume that all the individual downward thermal transition probabilities are equal, which is questionable, then they are roughly equal to $1/4T_1$ not $1/2T_1$, T_1 being the spin–lattice relaxation time between a pair of levels.

These criticisms, however, are not intended to detract from a very fine piece of pioneering experimental work but simply to indicate areas where the ground covered is less than firm.

The measurement of the noise temperature of the cavity maser described by McWhorter, Meyer and Strum again reflects the experimental skill of these workers. The method employed is straightforward but the accuracy obtained is high particularly when one reflects that the maser was operating under conditions of high gain (28 dB) and narrow bandwidth (60 kc/s) where the gain stability would be low. The significance of this experiment is that it confirmed for the first time the predictions of Bloembergen that the maser would have an equivalent noise temperature lower than that of any other known microwave amplifier.

PAPER 13*

Solid-State Maser Amplifier†

ALAN L. MCWHORTER and JAMES W. MEYER
Lincoln Laboratory, Massachusetts Institute of Technology, Lexington, Massachusetts
(Received August 14, 1957)

Summary

The operation of a solid-state maser amplifier at 2800 Mc/sec is described. A dual-frequency cavity containing paramagnetic potassium chromicyanide in an isomorphous cobalt diluent is used at 1.25°K. The experimental observations of the maser both as an amplifier and as an oscillator are compared with theory.

I. Introduction

A solid-state maser of the type proposed by Bloembergen[1] has been operated both as an amplifier and as an oscillator at 2800 Mc/sec, using $K_3Co(CN)_6$ containing 0.5% Cr as the paramagnetic salt. This material is particularly suited for maser application by virtue of its unusually long spin–lattice relaxation time, which was found to be 0.2 sec at 1.25°K by resonance saturation techniques.

The upper three of the four energy levels of the Cr^{+++} ion were used, with the energy level spacing suitably adjusted by means of the magnitude of the dc magnetic field and its orientation with

* *Phys. Rev.* **109,** 312–18 (1958).

† The research reported in this document was supported jointly by the Army, Navy, and Air Force under contract with Massachusetts Institute of Technology.

(1) N. Bloembergen, *Phys. Rev.* **104,** 324 (1956).

respect to the crystalline electric field of the salt. Spin state populations were inverted by saturating the resonance absorption at 9400 Mc/sec.

The amplifier, regenerative in nature, has as a result much narrower band width than one would expect from a casual consideration of the circuit Q's involved. For simplicity the measurements reported below were made with a relatively high Q reflection cavity. To achieve the large band widths inherent in the width of the paramagnetic resonance line without sacrificing gain, a very low Q structure, e.g., a slow wave structure, containing a much larger volume of the paramagnetic salt would have been necessary. In spite of its band-width limitation, however, the present circuit shows that a solid-state maser can be made to operate in reasonable agreement with the theoretical predictions.

II. Energy Levels

Baker, Bleaney and Bowers[(2)] have interpreted the paramagnetic resonance spectra of $K_3Cr(CN)_6$ as arising from two magnetically similar but differently oriented complexes per unit cell, with the spin Hamiltonian

$$H = \beta \mathbf{H} \cdot g \cdot \mathbf{S} + D[S_z^2 - \tfrac{1}{3}S(S+1)] + E(S_x^2 - S_y^2),$$

where for cobalt as the diluent, $D = 0.083$ cm^{-1}, $E = 0.011$ cm^{-1}, and g is approximately isotropic and equal to 1.99. The direction cosines between the magnetic axes (x, y, and z) and the pseudo-orthorhombic crystalline axes (a, b, and c) are given as

	x	y	z
a	0.104	0	0.994
b	± 0.994	0	∓ 0.104
c	0	1	0

Several energy level diagrams computed from this spin Hamil-

(2) Baker, Bleaney and Bowers, *Proc. Phys. Soc.* (*London*) **B69,** 1205 (1956).

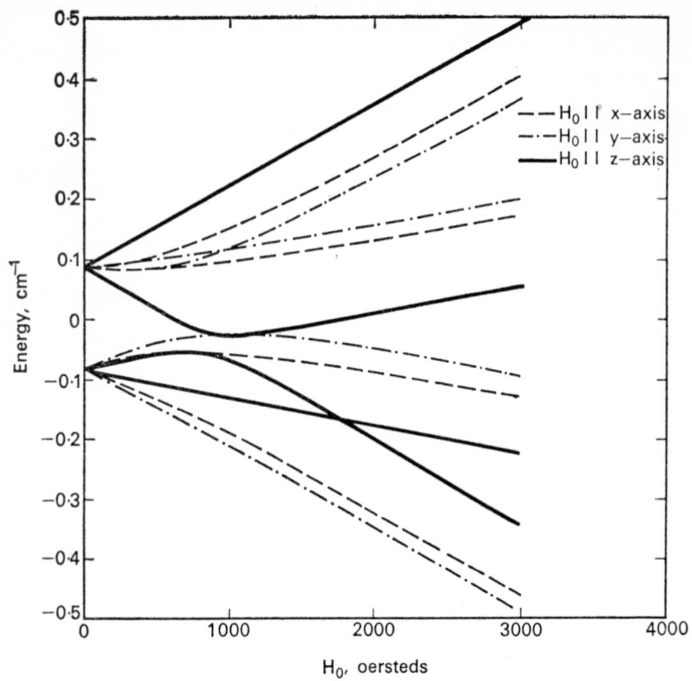

FIG. 1. Energy level diagram $K_3Cr(CN)_6$ with orientation of magnetic axes as parameters.

tonian are shown in Figs. 1 and 2. Within experimental accuracy, our preliminary measurements agree with these curves both in the 3 cm and 10 cm wavelength regions.

As these data indicate, several combinations of magnetic field and crystal orientation can be chosen which will permit operation at the selected frequencies, 2800 Mc/sec and 9400 Mc/sec. The combination which gave the best results in the initial investigation, and which was used for the measurements reported below, is shown in Fig. 3. The desired splitting was achieved by means of a slight rotation, about the *a* axis, of the *c* axis from the dc magnetic field.

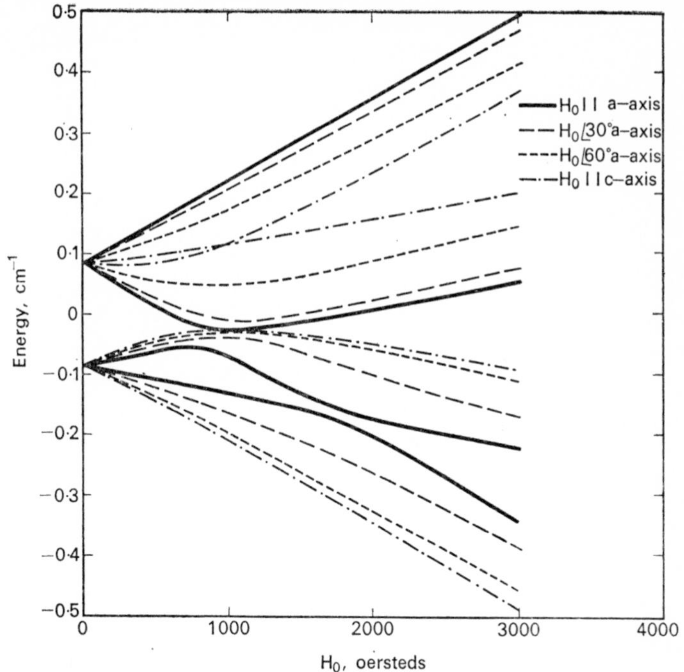

FIG. 2. Energy level diagram $K_3Cr(CN)_6$ rotation about *b* axis.

III. Apparatus and Method

A. Microwave apparatus

Because of its essential simplicity, a fixed-tuned, dual-mode coaxial microwave cavity was employed. Minor adjustment of the splitting of the energy levels to correspond to the resonant frequencies of the cavity was done by small shifts in crystal orientation. The cavity is one-half wavelength long at 2800 Mc/sec when operated in the *TEM* mode with the sample in place and with the remainder of the cavity volume filled with liquid helium. At 9400

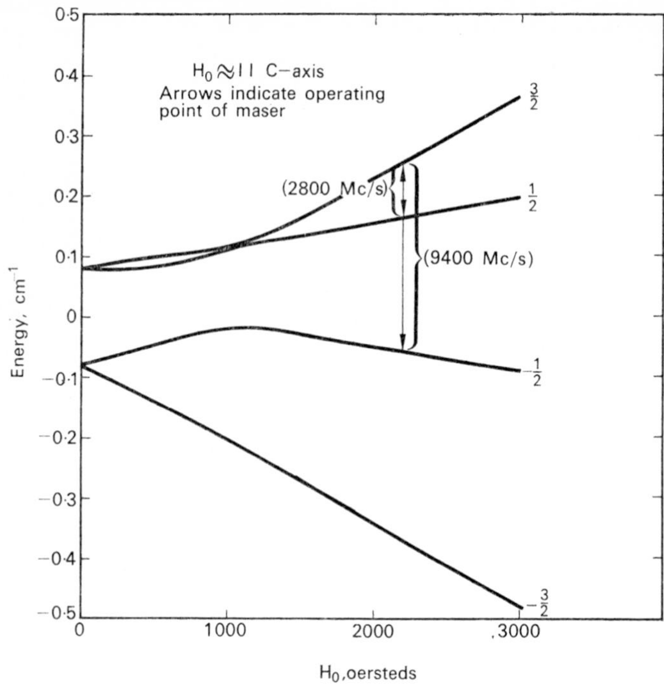

FIG. 3. Energy level diagram $K_3Cr(CN)_6$ showing point of maser operation.

Mc/sec, the cavity operated in the $TE_{11\,3}$ mode, which was especially chosen to reduce cross coupling of 9400-Mc/sec power into the coaxial line employed for 2800-Mc/sec operation. The cross coupling was further reduced by means of low-pass filters in the external coaxial line. The resulting magnetic field configurations and cavity features are shown in Fig. 4. Saturating power was coupled to the cavity by means of a section of silver-plated stainless steel wave guide terminated by a magnetic coupling hole. The size of this hole was adjusted to give approximately critical coupling when the paramagnetic resonance was saturated. The 2800-Mc/sec coaxial line was coupled to the cavity by means of a

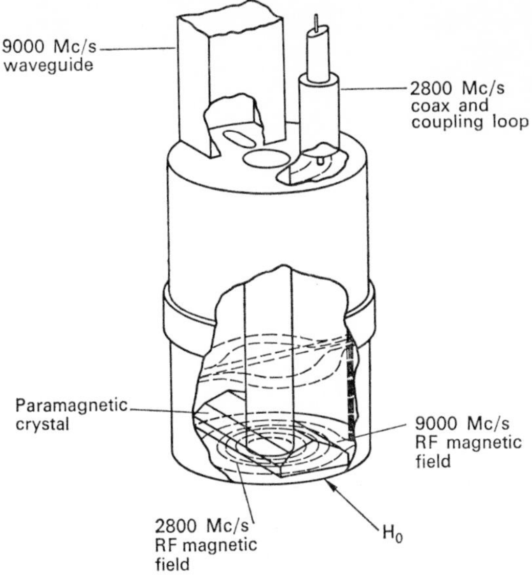

FIG. 4. Dual-frequency maser cavity.

loop. The extent of coupling could be adjusted both by rotation and depth of immersion of the loop in the cavity.

The low-temperature head, Fig. 5, was installed in a suitable double Dewar system. Thermal isolation of the cavity provided by the stainless-steel wave guide and coaxial line was sufficiently good to permit several hours of operation at 1.25°K with approximately one liter of liquid helium.

B. Sample preparation

Single crystals were grown from an aqueous solution of amounts of cobalt and chromium potassium cyanide appropriate to the chromium concentration desired. Crystals were prepared with chromium concentrations of from 0.1% to 2%. Standard crystal-growing techniques were employed to produce crystals of more

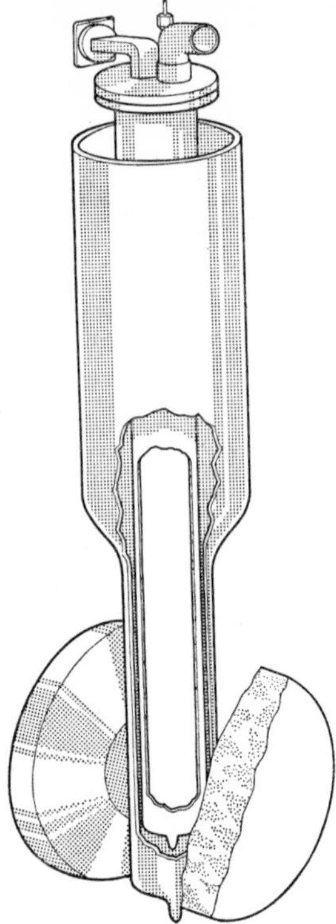

FIG. 5. Low-temperature head and double Dewar system.

than one square centimeter in cross section by three to five centimeters long.

C. Crystallography

The single crystals had two principal growth habits, as indicated in Fig. 6, one which produced a crystal form elongated in the c-axis direction with prominent m faces {110},[3] the other which produced flat hexagonal plates with prominent a faces {100}. The

[3] P. Groth, *Chemische Kristallographie* (W. Engelmann, Leipzig, 1906), Vol. 1, p. 422.

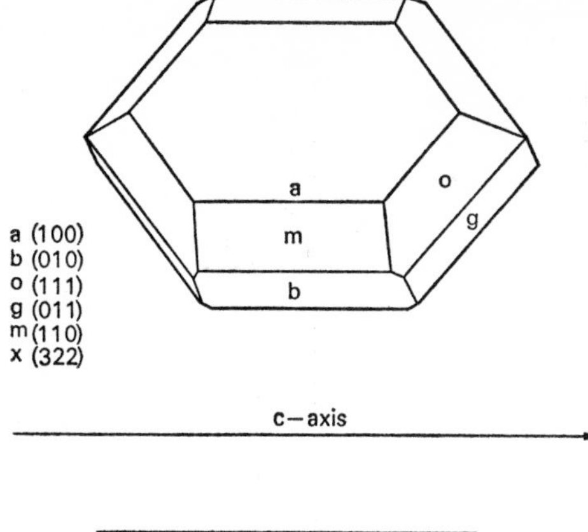

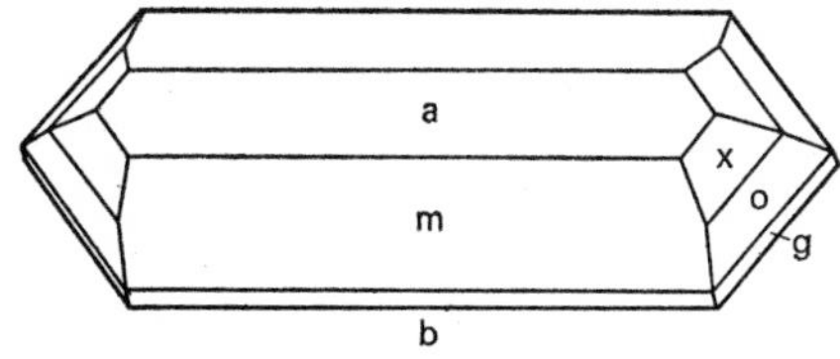

FIG. 6. Typical growth habit and crystal forms, $K_3Co(CN)_6$—$K_3Cr(CN)_6$.

correct orientation was determined from goniometric measurements on the well-formed faces.

IV. Experimental Results

A. Amplifier characteristics

The operation of the maser as an amplifier was investigated by applying the input power to the cavity through a directional coupler as illustrated in Fig. 7. This arrangement permitted gain–band-width measurements on the reflection cavity type amplifier through its single coaxial coupling line without the use of a circulator. The gain was determined by the amount of attenuation needed in the maser output line to maintain a constant signal

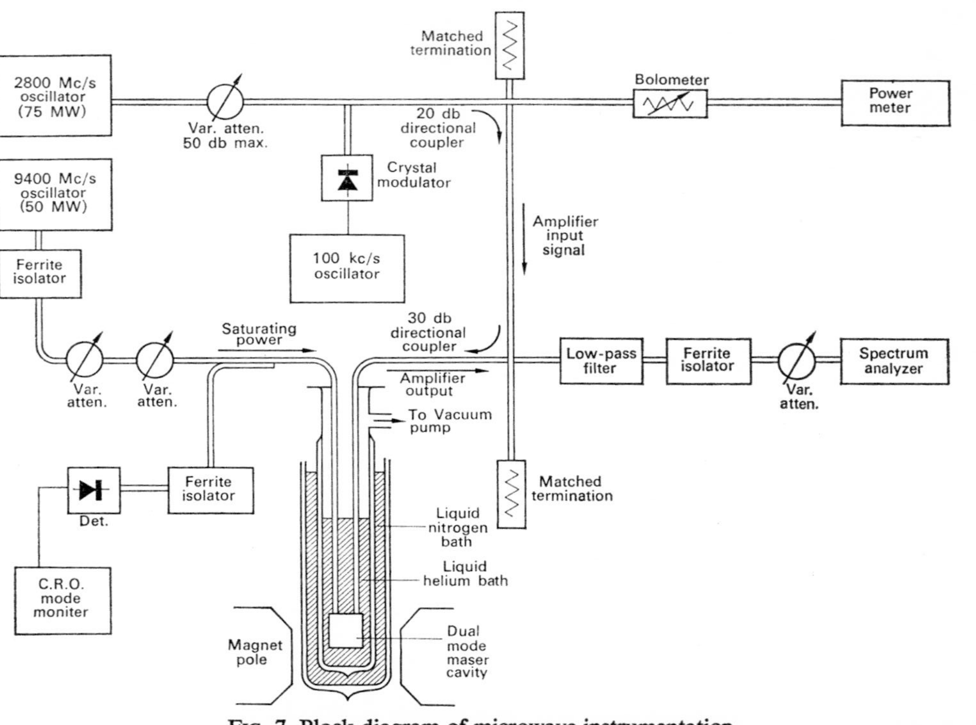

FIG. 7. Block diagram of microwave instrumentation.

amplitude at the spectrum analyzer. This additional attenuation in the output line, together with the ferrite isolator, served the additional purpose of keeping any power reflected from the spectrum analyzer from reaching the maser and being reamplified.

The band width was taken as the total frequency deviation required to reduce the amplifier power output to one-half its mid-band value. Band widths were measured on the spectrum analyzer after calibrating its frequency axis with the modulating scheme shown in the block diagram.

The results of the gain–band-width measurements are shown in Fig. 8. Parametric curves of both gain and band width are plotted as a function of 9400-Mc/sec power for two different values of 2800-Mc/sec external Q, which was adjusted by means of the degree of coupling. With still higher external Q it was possible to

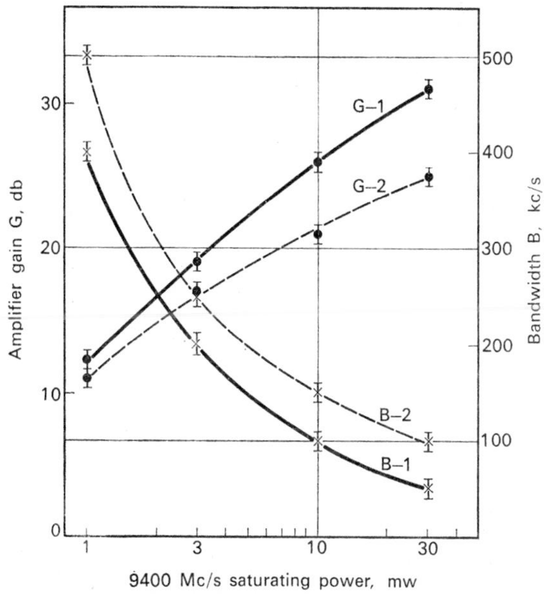

FIG. 8. Gain and band-width curves of maser amplifier with degree of coupling as a parameter.

achieve gains of 30 db or more with only 1 mw of saturating power, although the maser would then oscillate at the larger saturating powers. Stable gains of 37 db with 25 kc/sec band width were also possible. In all of these cases the band widths were limited by the Q of the associated circuitry and not by the intrinsic band width of the paramagnetic resonance, which here was in the 30–50 Mc/sec region.

Observations of the gain as a function of input 2800-Mc/sec power revealed the expected decreased gain as the difference in populations established by the saturating power is affected by the signal power. There was no change in gain when the signal power was increased from 10^{-11} to 10^{-10} watt, but thereafter the gain diminished and the band width increased.

B. Oscillator characteristics

The initial investigation of the maser as an oscillator was made by using a frequency-modulated probing signal applied to the coaxial coupling line. The frequency of the probing oscillator is swept by the time base of the oscilloscope and the power reflected from the cavity is displayed on the y axis as a function of frequency. In Fig. 9(*a*) we see the absorption resulting from the 2800-Mc/sec microwave resonance centered in the klystron mode pattern. With the magnetic field adjusted for paramagnetic resonance, the power reflected from the undercoupled cavity increases, [Fig. 9(*b*)]. The application of 9400-Mc/sec power [Fig. 9(*c*)] shows how the negative resistance produced by maser action improves the Q of the cavity, which in turn improves the coupling although no changes were made in the coupling loop adjustments. Further increase of saturating power enhances this effect [Fig. 9(*d*)], and in Fig. 9(*e*) the maser is beginning to produce power at 2800 Mc/sec. In Fig. 9(*f*) the beat signal between the output of the oscillating maser and the frequency-modulated probe signal is clearly seen with the video detector system.

The output of the oscillating maser was also observed on a spectrum analyzer in the absence of an input 2800-Mc/sec signal.

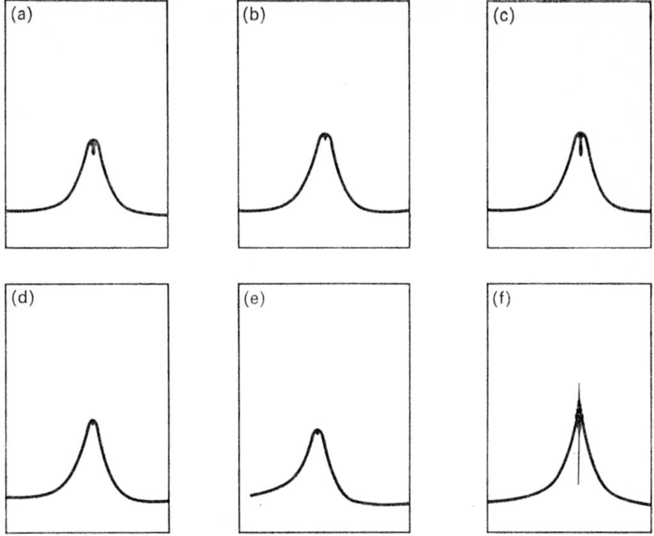

FIG. 9. Oscilloscope display of maser oscillator operation.

Maser power out as a function of saturating input power is shown in Fig. 10. The efficiency, P_0 (2800 Mc/sec)/P_i (9400 Mc/sec) is also given. The maximum efficiency obtained as operated here was -28.5 db, or 0.14%.

The realization of an oscillator with such small amounts of saturating power demonstrates the usefulness of chromium cobalticyanide as a maser material by reason of its advantageously long relaxation time. The relaxation time is not so long, on the other hand, that the maser will not handle reasonable signal input powers as an amplifier.

C. *Relaxation times*

The product of the phenomenological relaxation times, spin–lattice T_1 and spin–spin T_2, were measured using the power satura-

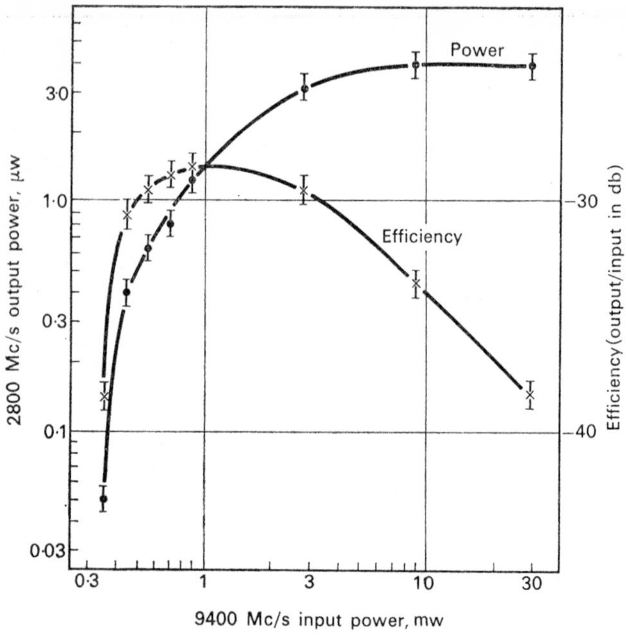

FIG. 10. Power output characteristics of maser oscillator.

tion of paramagnetic resonance technique.[(4)] Values for T_2 were then obtained from line width measurements, and values of T_1 subsequently obtained. Data were taken for the three crystalline axes, a, b, c, and the three $\Delta m_s = 1$ transitions at 1.25°K. Although the spin–spin relaxation time varied considerably the value of T_1 remained 0.2 second to within experimental error for all three transitions along the principal axes. T_2 meanwhile took on values from 4×10^{-9} sec to 9×10^{-9} sec and was largest for the $-\frac{1}{2} \rightarrow +\frac{1}{2}$ transition at all orientations. The sample contained 0.5% chromium concentration.

[(4)] Bloembergen, Purcell and Pound, *Phys. Rev.* **73,** 679 (1948); A. M. Portis, *Phys. Rev.* **91,** 1071 (1953); A. H. Eschenfelder and R. T. Weidner, *Phys. Rev.* **92,** 869 (1953).

V. Discussion

The gain of a reflection-cavity type maser amplifier, connected with directional couplers and ferrite isolators as was done here for measurement purposes, or with a circulator as would be used for the practical amplifier, is simply the ratio of the reflected to the incident power. In terms of the external and total cavity Q's, the gain can therefore be expressed as

$$G = [(1/Q_e - 1/Q_c)/(1/Q_e + 1/Q_c)]^2, \tag{1}$$

where

$$1/Q_c = 1/Q_u + 1/Q_M \tag{2}$$

gives the total cavity Q in terms of the unloaded Q and the magnetic Q resulting from the paramagnetic resonance. The magnetic Q is positive for resonance absorption and negative for stimulated emission, and the unloaded Q includes all other losses such as those in the cavity walls, and the dielectric losses in the salt. At low temperatures, Q_u will generally be so large that it can be neglected in the gain formula. (Our measured Q_u was 23 000 at 1.25°K.) When one makes this approximation and uses the absolute value of the magnetic Q, the gain is

$$G = [(Q_e + |Q_M|)/(Q_e - |Q_M|)]^2. \tag{3}$$

The band width of the amplifier is given by the operating frequency f divided by the loaded Q, which with the above approximation gives

$$B \doteq f(1/Q_e - 1/|Q_M|). \tag{4}$$

Consequently,

$$G^{\frac{1}{2}}B = f(Q_e + |Q_M|)/Q_e|Q_M|. \tag{5}$$

For a high-gain amplifier one adjusts the coupling so that Q_e almost equals $|Q_M|$. Under this condition

$$G^{\frac{1}{2}}B \doteq 2f/Q_e \doteq 2f/|Q_M|. \tag{6}$$

Hence, as was pointed out by Strandberg,[5] the square root of the gain times band width should be approximately a constant for a given volume of the salt in a given cavity configuration. Experimentally, this relation is obeyed very closely, the value being about 1.8×10^6 sec^{-1}. By decreasing the external Q, gain can be traded for band width, but at a considerable sacrifice of the former. Since the first part of Eq. (6) shows that the band width of the loaded cavity without the magnetic material is reduced by $G^{\frac{1}{2}}/2$ when it is operating as an amplifier, it is apparent that heavy coupling is required for even modest band width.

Equation (3) shows that the gain of a high-gain amplifier is very sensitive to small relative changes of Q_e and Q_M. As a result, slight changes in coupling or degree of saturation can change the gain and band width by large amounts even though $G^{\frac{1}{2}}B$ remains sensibly constant. This characteristic also indicates that the designer of a pulsed-type two-level maser will have to exercise considerable ingenuity to prevent wide excursions of gain and band width during the amplifying period.

An expected theoretical value for the $G^{\frac{1}{2}}B$ product can be calculated approximately in the following way: Expressing the magnetic Q in terms of the cavity volume V_c, the average rf magnetic field $\langle H_c^2 \rangle$, the operating frequency ν_{32}, and the magnetically absorbed or emitted power P_M, we have

$$1/\,|\,Q_M\,| = 4P_M/\nu_{32}\langle H_c^2\rangle V_c, \tag{7}$$

where, in Bloembergen's[1] notation,

$$P_M \doteq \frac{Nh^2\nu_{32}}{3kT}\left(\frac{w_{21}\nu_{21}-w_{32}\nu_{32}}{w_{32}+w_{21}}\right)W_{32} \tag{8}$$

in the approximation that one has full saturation at frequency ν_{13} and small power at the amplifying frequency ν_{32}. The formula for W_{32}, as given by Bloembergen, Purcell, and Pound,[4] is

$$W_{32} = \tfrac{1}{4}\gamma^2(H_1/2)^2 g(\nu)(S+M)(S-M+1). \tag{9}$$

[5] M. W. P. Strandberg, *Phys. Rev.* **106**, 617 (1957).

At the operating temperature of 1.25°K, the density of spins in the upper three levels of the chromium quartet is approximately $10^{19}/cm^3$. Assuming the w's are equal, as experimental results indicate, using approximately a 10% filling factor and the relation $g(\nu)_{max} = 2T_2$ ($T_2 = 6 \times 10^{-9}$ sec), we have for the $G^{\frac{1}{2}}B$ product a value 2.6×10^6 sec^{-1}, which is in reasonable agreement with the experimental value, 1.8×10^6 sec^{-1}.

It is also possible to compare the power output of the maser oscillator with the theoretically predicted output. Assuming full and homogeneous saturation of the resonance at ν_{31}, and a large signal amplitude at ν_{32}, the magnetic power is given by:[(1)]

$$P_M = (Nh^2\nu_{32}/3kT)(w_{21}\nu_{21} - w_{32}\nu_{32}). \qquad (10)$$

Taking $N = 3.9 \times 10^{19}$, the population of the upper three levels at 1.25°K, and $w_{21} = w_{32} \doteq 1/2T_1$, we obtain the value 8.7 microwatts, which is to be compared with the experimentally obtained four microwatts. Cavity and coupling losses will account for much of the discrepancy beteeen the two values.

It should be pointed out that a long spin–lattice relaxation time, which permits amplification with low saturating power, limits in a like manner the maximum output of the maser as an oscillator.

The operation of a solid state maser of this type places somewhat stringent requirements upon the paramagnetic material. At least three energy levels are necessary, and the paramagnetic ion must be in a field of sufficient asymmetry to produce mixed states and allow $\Delta m_s = 2$ quantum jumps. Not only should the interionic distances be great enough to reduce the spin–spin interaction to a point where saturation is practicable, but also there must be a sufficient number of spins participating in the maser action to overcome the cavity and coupling losses. Moreover, the residual orbital momentum should not be too great in order to assure a long spin–lattice relaxation time at low temperatures. Further, one would like substantial control over the separation of the energy levels by means other than changing the magnitude of the dc magnetic field (e.g., by varying the orientation of the magnetic field relative to the crystalline electric field), in order to have more freedom in the choice of both saturating and operating frequencies. These requirements severely limit the choise of a maser salt among paramagnetic substances with characteristics reported in the literature.[6, 7] Of the two working substances investigated in detail by us, zinc fluosilicate containing a small percentage of Ni^{++} and potassium cobalticyanide containing a small percentage of Cr^{+++}, only the latter could be used for successful operation of the maser described here.

The other paramagnetic salt mentioned by Bloembergen, gadolinium ethyl sulfate, was not tried because it offered much less flexibility of choice of saturating and operating frequencies. Moreover, as outlined by Scovil, Feher and Seidel,[8] the relaxation time between the upper two levels had to be drastically altered in order to produce oscillations. This further limits the frequency flexibility. Because of previous experience with the salt,[9] zinc fluosilicate containing a small percentage of Ni^{++} was first tried by us in a maser designed to operate at 1400 Mc/sec while saturating at 9000 Mc/sec. We were unable to get this maser to operate for two main reasons: first, because the spin–lattice relaxation time is of the order of 10^{-4} sec, considerable power is required to saturate the resonance; secondly, the line width in the diluted salt was inhomogeneously broadened by a distribution of crystalline electric fields. As a result of the inhomogeneous broadening only $T_2{}^*/T_2$ of the paramagnetic spin concentration was available for participation in the maser action, where $T_2{}^*$ is the time associated with the total width of the inhomogeneous line.

An attempt to improve the probability of operation by going to a higher amplifying frequency, 2800 Mc/sec, was also unsuccessful.

(6) B. Bleaney and K. W. H. Stevens, *Reports on Progress in Physics* (The Physical Society, London, 1953), Vol. 16, p. 108.

(7) K. D. Bowers and J. Owen, *Reports on Progress in Physics* (The Physical Society, London, 1955), Vol. 18, p. 304.

(8) Scovil, Feher and Seidel, *Phys. Rev.* **105,** 762 (1957).

(9) J. W. Meyer, Ph.D. thesis, University of Wisconsin, 1955 (unpublished).

The nickel ion is still attractive, however, from other points of view. Because it is a spin triplet without nuclear magnetic moment, ions need not be sacrificed by distribution over unused energy levels. Its integral spin $S = 1$ leaves one level, $m_s = 0$, unchanged by the application of the external magnetic field, thereby enhancing its tunability. Further work is in progress to find a suitable salt and diluent for the nickel ion, as part of a more general program which includes the paramagnetic resonance spectroscopy of other materials.

Published paramagnetic resonance data reveal few salts which appear to be suitable materials for maser operation. The cyanides were especially attractive because the reduction of line width by dilution is not limited by the proton magnetic moment as it is in the hydrated salts. In the latter case, lines of a few oersteds width can only be achieved by deuteration. Moreover, the spin–lattice relaxation time is long, which when combined with a reduced spin–spin interaction requires little power for saturation of the resonance.

Thus far, no experimental measurement of the noise figure of this maser amplifier has been made, but work is in progress to evaluate this most important characteristic.†

VI. Acknowledgments

We should like to thank Professor Bloembergen for disclosing his concept of a solid-state maser prior to its publication, and for subsequent discussions of related problems. The assistance of Dr. S. H. Autler with some of the experimental work is also gratefully acknowledged. We are indebted to Dr. Harry C. Gatos and Mr. L. J. Gordon of this laboratory, and Dr. C. W. Wolfe of Boston University for furnishing the single crystals of potassium cobalticyanide containing chromium. The physical preparation and orientation of the samples was done by Dr. S. A. Kulin, and Mr.

† *Note added in proof.*—A preliminary sequence of measurements, giving an upper limit of 20°K for the noise temperature of the maser, has now been completed: McWhorter, Meyer and Strum, *Phys. Rev.* **108,** 1642 (1957).

E. P. Warekois. Assistance in the mechanical design of the cavities was provided by Mr. A. M. Rich. Finally, we should like to acknowledge profitable discussions with Dr. R. H. Kingston and Dr. H. J. Zeiger and the encouragement and continued support of Dr. Benjamin Lax.

PAPER 14*

Noise Temperature Measurement on a Solid-State Maser†

A. L. McWhorter and J. W. Meyer

Lincoln Laboratory, Massachusetts Institute of Technology, Lexington, Massachusetts

and

P. D. Strum

Ewen Knight Corporation, Needham, Massachusetts

(Received October 14, 1957)

Noise measurements have been made on a three-level,[1] 2800-Mc/sec solid state maser with sufficient accuracy to establish that its noise temperature does not exceed 20°K. The maser was operated as a reflection cavity amplifier at 1.25°K, with 9000-Mc/sec saturating power, and used $K_3Co_{0.995}Cr_{0.005}(CN)_6$ as the mixed paramagnetic salt. The amplifier and oscillator characteristics are the subject of a forthcoming article.[2]

The maser noise temperature T_M was determined by measuring the noise outputs of the system, N_1 and N_2, corresponding to two noise inputs with temperatures T_1 and T_2, respectively. The output noise N_1 is proportional to $(T_1+T_M)G_M+T_R$, and N_2 is proportional to $(T_2+T_M)G_M+T_R$, where G_M is the gain of the maser and

* *Phys. Rev.* **108,** 1642–4 (1957).

† The research reported in this document was supported jointly by the Army, Navy, and Air Force under contract with the Massachusetts Institute of Technology.

(1) N. Bloembergen, *Phys. Rev.* **104,** 324 (1956).

(2) A. L. McWhorter and J. W. Meyer, *Phys. Rev.* (to be published).

T_R is the noise temperature of the receiver following the maser. T_M is then given by

$$T_M = (T_2 - T_1)/(N_2/N_1 - 1) - T_1 - T_R/G_M. \quad (1)$$

Figure 1 shows the experimental arrangement which was used. Temperature T_1 corresponded to the noise input with the argon discharge tube off, so that it was primarily the noise generated by

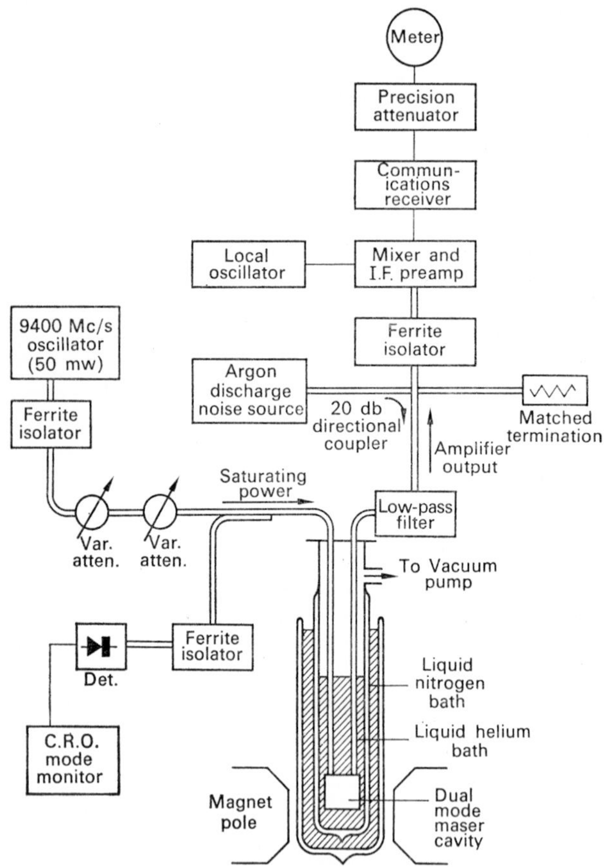

FIG. 1. Block diagram of microwave instrumentation.

the room temperature (30°C) ferrite isolator, while T_2 was the total noise input with the argon tube energized. With 15 milliwatts of 9000-Mc/sec pumping power, enough to insure full saturation but not enough to cause heating of the salt, the external coupling at the amplifying frequency was adjusted to give a maser gain of 28 db with about 60-kc/sec band width. A communications receiver was used to provide a receiver band width small in comparison with that of the maser. The entire receiver had a noise temperature of about 1800°K (8.5 db noise figure). As a result of many measurements made with the precision attenuator by two

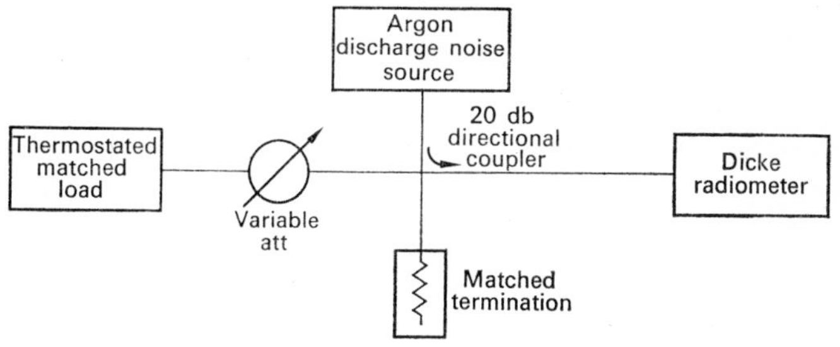

FIG. 2. Block diagram of calibration measurement.

different operators, the ratio N_2/N_1 was determined to be 1.125 ± 0.025 db.

The output of the argon discharge noise source through the directional coupler was calibrated by using a matched load, thermostated in an oil bath to ± 0.1°C, and a Dicke radiometer in the arrangement shown in Fig. 2. The temperature of the thermostated load was adjusted until the radiometer output obtained with the variable attenuator set at zero and the argon tube off, equalled the output obtained with the attenuator fully in and the argon tube on. Knowing the temperature of the thermostated load and of the variable attenuator, one could then determine the noise output of the discharge tube by this essentially null type of measurement. At the frequency of the maser, the noise from the discharge tube through the coupler was 99.5 ± 1.5°K.

In order to establish the noise temperature of the maser itself, T_1 and T_2 must be corrected for the attenuation in the input line and the thermal noise generated in this line. The input circuit after the directional coupler consisted of a wave-guide-to-coax coupling, a low-pass filter, and a section of coaxial cable, having a combined attenuation of 0.15 db at room temperature, followed by a 27-inch section of silver-plated stainless steel coaxial line leading down through the helium Dewar to the maser cavity. This line had an attenuation of 0.7 db at room temperature and 0.4 db under operating conditions. Since it was not known what temperature distribution corresponded to the 0.4-db attenuation, two extreme cases were calculated for the noise generated by the line. The most unfavorable case, from the standpoint of a low calculated value for T_M, is to assume that the entire attenuation takes place at 0°K. The most favorable case, consistent with the physical situation, was assumed to be a uniform temperature gradient over the entire length of the coaxial line, with the attenuation varying as the square root of the temperature. Upon using $N_2/N_1 = 1.125$ db, and 99.5°K for the noise from the argon tube, Eq. (1) gives for the first case $T_M = 8$°K, while for the second, $T_M = -8$°K. When the uncertainties in the other quantities are taken into consideration, the calculated value of T_M is 0 ± 19°K. Hence an upper limit of about 20°K can be established for the maser noise temperature.

If one neglects the small thermal radiation from the cavity walls, spontaneous emission at the amplifying frequency should be the only source of noise in the maser.[(3)] Using Bloembergen's notation,[(1)] the theoretical noise temperature under fully saturated conditions would then be

$$T_M = T_B \frac{\nu_{32}(w_{21}+w_{32})}{w_{21}\nu_{21}-w_{32}\nu_{32}},$$

where T_B is the helium bath temperature. Assuming the w's are equal, we have $T_M = 2$°K.

(3) R. V. Pound, *Ann. Phys.* **1**, 24 (1957); M. W. Muller, *Phys. Rev.* **106**, 8 (1957); M. W. P. Strandberg, *Phys. Rev.* **106**, 617 (1957); Shimoda, Takahasi and Townes, *J. Phys. Soc. Japan* **12**, 686 (1957).

The 20°K uncertainty is a result of three main contributions, all approximately equal: the measurement of $(N_2/N_1 - 1)$, the calibration of the argon discharge noise source, and the uncertainty in the noise generated by the input line. To obtain the best accuracy in the measurement of T_M, one should use noise inputs comparable with the maser noise, but the present necessity of using a room-temperature isolator, which gives a minimum noise input of about 300°K, prevented this. Work is now in progress to develop a good room-temperature circulator or a low-temperature isolator. Such circuit elements, and the subsequent use of low-temperature noise sources, would also permit the generation of noise temperatures T_1 and T_2 by thermostated matched loads at accurately known temperatures, thus eliminating the calibration of a secondary standard. In the masers now being designed, the replacement of the coaxial input line with a wave guide will eliminate most of the input attenuation and hence the uncertainty arising from noise generation in the input circuit. These improvements should reduce the uncertainty in the measurement of T_M to about 5°K.

We wish to thank Professor L. D. Smullin for the use of his Dicke radiometer in the calibration of the argon tube.

Comments on Paper 15

WE HAVE spent some time looking at the theory of the T.W.M. because it seemed to be desirable to establish a common point of view for the T.W.M. and cavity devices. The first published description of a working T.W.M. is contained in the paper of De Grasse, Schulz-du Bois and Scovil which is reproduced in full in the following section. The early part, in particular Section II, is taken up with a discussion of the theory of the T.W.M. which is essentially similar to the discussion of Chapter 3. To relate the B.T.L. results to those quoted earlier the reader should note that

$$F_{+} = \frac{\eta}{(1+R)}$$

and pay due regard to the fact that m.k.s. units are used in the B.T.L. paper whereas we have quoted Q_m in c.g.s. units

$$\left(\frac{1}{Q_m}\,(\text{m.k.s.}) = \frac{1}{Q_m}\,(\text{c.g.s.}) \times \frac{1}{4\pi}\right)$$

There is also an error in this part of the Bell paper, the transition rate under circularly polarised excitation quoted (equation (9)) is too small by a factor of 2. The performance of their maser material is thus underestimated by a factor of 2 and on the basis of their device performance they overestimate their filling factor by the same factor.

Section III of the paper is devoted to a discussion of slow wave structures suitable for the T.W.M. The need for structures with high slowing factors coupled with substantially circularly polarised r.f. magnetic fields is emphasised and it is rightly noted that structures consisting essentially of arrays of parallel conductors can satisfy both requirements and, furthermore, allow the pump power to propagate (not in a slow mode but this is not significant). The r.f. field distribution for the slow mode in such a structure approximates very closely to that of T.E.M. waves propagating in the *direction of the extension of the conductors*, there being a constant (at a given frequency) phase difference between the fields associated with successive conductors in the array. This is the basis of the method of analysis of such structures proposed by Fletcher (1952). Applying Fletcher's method to the comb structure gives for the equation determining the relationship between the frequency and phase constant (φ)

$$\tan kl = \frac{Y(\varphi)}{\omega C_1}$$

where

$k = \frac{\omega}{C}$; $C =$ velocity of propagation of T.E.M. waves along the conductors,

l = the length of each conductor in the array,

C_1 = finger and fringe capacity as defined by De Grasse *et al.*,

and

$Y(\varphi) =$ the characteristic admittance of each conductor in the array.

The calculation of $(Y\varphi)$ has been discussed by Fletcher, Mourier and Leblond and Walling (all quoted in the B.T.L. paper). The method adopted by De Grasse is essentially that of Mourier and Leblond who give as a first approximation to $Y(\varphi)$

$$Y(\varphi) = \frac{1}{\sqrt{(\varepsilon\mu)}}\left[\gamma_p + 4\gamma_1 \sin^2 \frac{\varphi}{2}\right]$$

γ_p being the conductor to base plate capacity and γ_1 the capacity between adjacent conductors (both per unit length). The Bell Laboratories workers resorted to experiment to determine the various capacitances involved, i.e.

$$\gamma_p = \frac{\sqrt{(\varepsilon\mu)}}{Z_{01}} \quad \text{and} \quad \gamma_1 = \frac{\sqrt{(\varepsilon\mu)}}{Z_{02}}.$$

It should perhaps be pointed out that equation (35) of the paper appears to be incorrectly stated and should read

$$2\pi f C_1 \tan \frac{\pi f}{2f_0} = \left(\frac{1}{Z_{01}} + \frac{4}{Z_{02}} \sin^2 \frac{\varphi}{2}\right)$$

As the reader will readily see it is this equation which is solved by employing Fig. 9 of the B.T.L. paper, *not* their equation (35).

Perhaps the most important question in the design of slow wave structures for travelling wave masers is that of the effect of the heavy dielectric loading which is introduced by the maser material. This matter is touched upon in the paper under discussion and we may note that if the dielectric (maser material) is introduced in the region between the conductors and base plates (or side walls), as it usually is, then when $\varphi = 0$ the electric fields are entirely contained in this region and the effect of the dielectric is very marked. On the other hand, as φ increases an increasing fraction of the total electrical energy is stored between the conductors and the effect of the dielectric loading becomes less marked. For this reason structures with heavy loading tend to propagate over a wider range of frequencies (pass band) than the unloaded structure. As noted by the B.T.L. workers if the dielectric does not extend for the full distance between the conductors and the side walls the width of the pass band can be markedly reduced and the slowing factor C/V_g correspondingly increased. ($V_g = p(d\omega/d\varphi)$, where p = period of the structure.) This procedure though can result in the propagation of anomalous modes (two values of the phase constant for a given frequency) which are disastrous from the point of view of maser operation (De Grasse *et al.*, 1961).

The measurements on the gadolinium ethyl sulphate T.W.M. are of interest: in particular the remarkably high output power, +15 dBm (with 3 dB gain compression), which is quoted. Despite the good saturation characteristic and high gain per unit length which are reported, gadolinium ethyl sulphate is not a practical maser material—it has to be encased in a plastic container to avoid loss of water of crystallisation. Ruby is a much more practical material and the results quoted in Section VI of the paper paved the way for the subsequent development of the T.W.M. The problem of making the resonant fields of the isolator and maser material is solved by the B.T.L. workers in this paper, not by employing ferrite or garnet isolation for the isolator but by employing a concentrated ruby sample. This material tends to absorb pump power but the pump transition cannot be saturated except at very high pump input levels and the population of the signal transition levels are not inverted, thus resonant absorption occurs in the isolator ruby under the same conditions as resonance emission in the active low concentration ruby.

The noise temperature of the ruby T.W.M. was carefully determined, and completely supports theoretical expectations. It may be noted that because of the much greater bandwidth and stability of the T.W.M. when compared with the cavity device, accurate noise measurements are easier in the case of the T.W.M. The accuracy reported by the Bell workers was nevertheless a notable achievement at the time.

To conclude, although this paper contained a number of trifling errors it nevertheless describes a remarkable piece of research and device development which set the pattern for the many developments of T.W. masers which followed.

PAPER 15*

The Three-Level Solid State Traveling-Wave Maser†

R. W. DeGrasse, E. O. Schulz-Du Bois and H. E. D. Scovil

(Manuscript received December 18, 1958)

Summary

Broadband very-low-noise microwave amplification can be obtained from solid state maser action in a propagating microwave structure. Such a traveling-wave maser produces unilateral amplification with a high degree of gain stability. The theory of the traveling-wave maser is developed and used to compare the gain, bandwidth and gain stability of the traveling-wave maser with that of the cavity maser. The general requirements for traveling-wave maser slow-wave structures are discussed. Theoretical analysis and experimental results are presented for the comb-in-waveguide slow-wave structure.

A traveling-wave maser consisting of a ruby-loaded comb structure was tested. A gain of 23 db at 6 kmc with a bandwidth of 25 mc was obtained. Further performance characteristics of this amplifier and one using gadolinium ethyl sulfate are given. Experimental verification of the low noise temperature of solid state masers was obtained.

I. Introduction

The three-level solid state maser, as proposed by Bloembergen,[(1)] employs a microwave pump signal to alter the thermal equilibrium of a paramagnetic salt in such a manner that an otherwise absorptive medium becomes emissive when stimulated by radiation at the signal frequency. Successful application of this principle to produce microwave amplification was reported by

* *Bell Syst. Tech. J.* **38**, 305–34 (1959).

† This work was supported in part by the U.S. Army Signal Corps under Contract DA–36–039 sc–73224.

Scovil *et al.*,[2, 3] who used microwave cavities to couple the microwave radiation to the paramagnetic salt. Several laboratories[4, 5] have since operated such cavity-type masers.

Microwave amplification can also be obtained by stimulating radiation from active material in a propagating structure. Effective coupling of the microwave fields to the paramagnetic salt is obtained by slowing the velocity of propagation of the microwave energy through the structure. The active material produces an equivalent negative resistance in the slow-wave structure, and a propagating wave having an exponentially increasing amplitude is obtained.

However, if a slow-wave structure is simply filled with the active material, the device will be reciprocal and have gain in both directions. It would therefore require excellent input and output matches and presumably external isolation to obtain unilateral gain. Unidirectional traveling-wave amplification can be obtained in a slow-wave structure which has definite regions of circular polarization of the magnetic field if a maser material is employed which has circularly polarized signal-frequency transitions. The maser material is then loaded in the structure in such a manner that it is coupled to the structure only for one direction of wave propagation. Then, the maser will have high gain in one direction and little or no gain in the reverse direction.

Unidirectional gain alone is not enough to ensure freedom from regenerative instability. One might add reciprocal loss, as in the case of the traveling-wave tube. However, the traveling-wave tube has an electron beam to carry the microwave energy past the loss region, with very little forward attenuation; in the case of the traveling-wave maser, only the one propagation wave is present and the attenuation subtracts directly from the forward gain. Unidirectional loss may be obtained in the same manner as the unidirectional gain, that is, by excitation of a magnetic material through a circularly polarized magnetic field. For the magnetic material, one may use either an absorptive ferrimagnetic material or an absorptive material whose thermal equilibrium is not disturbed by the microwave pump power.

II. Traveling-Wave Maser Theory

The amplification in a traveling-wave maser is obtained from power transferred to the microwave circuit by coherent excitation of the paramagnetic spins due to the magnetic field of the circuit. This effect can be represented classically in the constitutive equation of electromagnetic theory†

$$B = \mu_0(H+M_m), \tag{1}$$

where μ_0 is the permeability of free space and M_m is the magnetic moment per unit volume of maser material. The magnetic moment, M_m, is computed from the quantum mechanical treatment of the paramagnetic spin system. From this analysis may be obtained a complex permeability, given by

$$\chi' - j\chi'' = \frac{M_m}{H}. \tag{2}$$

The real part of the permeability, χ', will produce reactive effects in the microwave circuit, while the imaginary part can produce gain or loss. In most cases, the magnetic permeability depends upon the orientation of the magnetic field, H; thus, it is in general a tensor quantity.

If we have a microwave structure uniform in the z direction and partially filled with a maser material, the rate of change of power in the circuit with distance is given by

$$\frac{dP}{dz} = -\frac{1}{2}\,\omega\mu_0 \int_{A_m} H \cdot \chi'' \cdot H^*\, dS, \tag{3}$$

where the integration is performed over the cross section of the maser material, A_m. The power in the waveguide is given by

$$P = \frac{1}{2}\, v_g \mu_0 \int_{A_s} H^2\, dS, \tag{4}$$

† All equations are given in MKS units.

where v_g is the group velocity in the waveguide circuit, and the integration is performed over the entire waveguide cross section, A_s. The gain in a length of structure, l, is then given by

$$G = \frac{P(l)}{P(0)} = \exp\left[-\chi''_{\max} F(\omega/v_g)l\right], \tag{5}$$

where $\chi''_{\max}$ is the magnitude of the diagonalized χ'' tensor, and the filling factor, F, is defined by

$$F \equiv \frac{\int_{A_m} H \cdot \chi'' \cdot H^* \, dS}{\chi''_{\max} \int_{A_s} H^2 \, dS}. \tag{6}$$

Rather than use the exact tensor representation for χ'', it is convenient to obtain a value for χ'' which reflects the magnetic field orientation. In a cavity maser, χ'' would be defined for linear polarization. In a traveling-wave maser, however, it is desirable to obtain nonreciprocal effects by using circularly polarized magnetic fields to excite the signal transition in the maser material.

In a maser material with small zero field splitting, such as gadolinium ethyl sulfate, the signal transitions have nearly pure circular polarization. Thus, if the magnetic fields of the circuit are resolved into circular polarized components, H_+ and H_-, the filling factor and gain (in decibels) are given by

$$F_+ = \frac{\int_{A_m} H_+{}^2 \, dS}{\int_{A_s} (H_+{}^2 + H_-{}^2) \, dS}, \tag{7}$$

$$G_{db} = -27.3 \chi''_+ F_+ \frac{fl}{v_g}, \tag{8}$$

where f is the signal frequency.

The permeability for positive circular polarization, χ''_+, can be calculated using the notation of Schulz-Du Bois.[(6)] He obtains for the rate of transition per ion from a state $\bar{n}$ to a state $\bar{n}'$,

$$w_{\bar{n}\to\bar{n}'}(S_+) = \frac{1}{2}\left(\frac{\pi g\beta}{h}\right)^{2} g(f-f_0) \mid \langle \bar{n}' \mid S_+ \mid \bar{n} \rangle \mid^2 H_+{}^2, \tag{9}$$

where positive circular polarization is assumed as indicated by S_+. The power absorbed per unit volume of material is then

$$P = (\rho_{\bar{n}} - \rho_{\bar{n}'}) hf w_{\bar{n}\to\bar{n}'}(S_+), \tag{10}$$

where $\rho_{\bar{n}}$ is the density of ions in energy state $\bar{n}$ per unit volume.

The power absorbed per unit volume is given classically in terms of χ''_+ as

$$P = \tfrac{1}{2}\omega\mu_0\chi''_+ H_+^2. \tag{11}$$

Thus, χ''_+ is given by

$$\chi''_+ = \frac{\pi}{2\mu_0 h}(g\beta)^2(\rho_{\bar{n}} - \rho_{\bar{n}'})g(f-f_0)\,|\,\langle \bar{n}' \,|\, S_+ \,|\, \bar{n}\rangle\,|^2. \tag{12}$$

If the equilibrium spin populations are inverted by making $\rho_{\bar{n}'}$ greater than $\rho_{\bar{n}}$, χ''_+ will be negative and amplification is obtained. The various conditions for maser population inversion have been discussed by Scovil.[(3)] A companion paper in this issue[(7)] discusses this problem in more detail.

For the case of propagation of energy in the reverse direction through the amplifier, the magnetic field sense of polarization will reverse and the filling factor will become

$$F_- = \frac{\int_{A_m} H_-^2\, dS}{\int_{A_s} (H_+^2 + H_-^2)\, dS}. \tag{13}$$

The degree of nonreciprocity of gain is then determined by the ratio, R_m, given by

$$R_m = \frac{\int_{A_m} H_+^2\, dS}{\int_{A_m} H_-^2\, dS}. \tag{14}$$

This ratio must be optimized in the selection of a suitable slow-wave structure for use in the amplifier.

As was mentioned before, nonreciprocity in gain is not enough to insure that the amplifier will not have regenerative effects due to mismatched input and output terminations. Thus, it is necessary to also include some nonreciprocal loss for isolation. Both ferri-magnetic and paramagnetic isolators have been used. By increasing the concentration of the active ion in the maser crystal, it is possible to prevent maser action. Such a high-concentration crystal will have energy levels which are identical to the maser material; thus, it will provide nonreciprocal loss at the desired magnetic field and orientation. Ferrimagnetic isolators can be designed to operate at the magnetic field required by the maser material by using shape anisotropy to change the frequency of ferrimagnetic resonance.

The isolator loss is determined by (5). If we retain the convention that H_+ produces excitation of magnetic spins for propagation in the direction of amplification, and H_- for the reverse direction, then the isolator loss in the reverse direction is given by

$$L = 27.3\chi''_{i+}F_{i-}\frac{fl}{v_g}, \tag{15}$$

where

$$F_{i-} = \frac{\int_{A_i} H_-^2\, dS}{\int_{A_s} (H_-^2 + H_+^2)\, dS} \tag{16}$$

and χ''_{i+} is the lossy permeability of the isolator. The ratio of reverse loss to forward loss, R_i, then is the figure of merit of the isolator:

$$R_i = \frac{\int_{A_i} H_-^2\, dS}{\int_{A_i} H_+^2\, dS}. \tag{17}$$

In general, the figure of merit of a paramagnetic isolator will be about the same as that of the paramagnetic maser material. Since χ''_i is usually much greater for a ferrimagnetic than for a paramagnetic, it is possible to locate a ferrimagnetic isolator of small cross section in a region which will optimize R_i.

Some loss, L_0, will occur in the TWM structure due to the usual resistive losses in the conductors. This loss is reduced below the usual room temperature value by a factor of 2 to 4 by operation at liquid helium temperatures. In some cases this insertion loss can be quite high, and care must be exercised in the selection of a circuit. Uniform current distribution and the largest possible surface area in the circuit conductors are desirable.

In discussing TWM circuits, it is useful to rewrite the factor

fl/v_g as the product of the slowing factor, S, and the number of free space wavelengths in the length of the structure, N, where

$$S = \frac{c}{v_g}, \qquad N = \frac{l}{c/f}. \tag{18}$$

Then the over-all maser forward gain and reverse loss equations are

$$G = 27.3SN\left(-\chi''_+F_+ - \chi''_{i+}\frac{F_{i-}}{R_i}\right) - L_0, \tag{19}$$

$$L = 27.3SN\left(\chi''_+\frac{F_+}{R_m} + \chi''_{i+}F_{i-}\right) + L_0. \tag{20}$$

For short-circuit stability of the amplifier, L must exceed G. In some cases, it may even be desirable to have L as much as 20 db greater than G, in order to eliminate any regenerative gain effects due to load changes. It is often convenient to refer to the product χ''_+F_+ as the inverse of the magnetic Q of the maser material. The magnetic Q will be defined for amplification in the forward direction as a positive number,

$$Q_m = \frac{-1}{\chi''_+F_+}. \tag{21}$$

Thus, the gain of the maser material only is simply

$$G = 27.3\frac{SN}{Q_m}. \tag{22}$$

The frequency variation of the TWM gain is given primarily by the term $g(f-f_0)$ in (12). If a Lorentzian line shape is assumed for the maser material, then

$$g(f-f_0) = \frac{2}{\pi B_m}\frac{1}{1+\left(2\dfrac{f-f_0}{B_m}\right)^2}, \tag{23}$$

where B_m is the bandwidth over which χ'' is greater than one half its peak value. If we assume that the permeability is an analytic function, then it follows that, for the Lorentzian line shape,

$$\chi'_{+} = -2\frac{f-f_0}{B_m}\chi''_{+}. \tag{24}$$

This rapid variation of χ'_{+} in the vicinity of the amplifying region will produce some perturbation in the phase velocity characteristics of the slow-wave structure. However, in a broadband maser slow-wave structure this effect will be negligible. Using the frequency variation of (23), we obtain for the 3-db bandwidth of a traveling-wave maser

$$B = B_m\sqrt{\frac{3}{G_{db}-3}}. \tag{25}$$

This derivation of bandwidth assumes a Lorentzian line shape. The actual emission line shape of a maser material depends upon a number of factors, and, at the present state of the art, is best determined experimentally.

The bandwidth variation given by (25) is quite different from that predicted for the cavity maser and it is apparent that the gain–bandwidth product increases at high gain, rather than reaching a constant as in the case of the cavity maser (CM).

The bandwidth variation with gain has been plotted in Fig. 1 for a traveling-wave and a cavity maser. The cavity maser was assumed to have a magnetic Q equal to that of the TWM. For a typical case, we may take 0.05 per cent Cr^{+++} in Al_2O_3, for which B_m is 60 mc for operation at 6 kmc with the magnetic field at 90° to the crystal axis. In this operation, a magnetic Q of 150 is obtainable at 1.5°K. The gain of a cavity maser has been calculated, taking into account the effect of χ' from (24), as

$${G_{\text{CM}}}^{1/2} = \frac{(1-jb_0)(1+jb_m)-R_0/R_m}{(1+jb_0)(1+jb_m)+R_0/R_m}, \tag{26}$$

where

$$b_0 = \left|\frac{R_0}{R_m}\right|\frac{2\Delta f}{f_0}Q_m, \tag{27}$$

$$b_m = 2\frac{\Delta f}{B_m}, \tag{28}$$

R_0 = the effective load impedance of the circulator,

R_m = the effective resistance of the maser material

and

Q_m = the magnetic Q of the maser material.

In the limit of large gain, the gain–bandwidth product approaches a constant given by (26) as

$$G_{CM}{}^{1/2}B = \frac{2}{\left(\frac{Q_m}{f_0}+\frac{1}{B_m}\right)}. \tag{29}$$

For the assumed maser operation in ruby,

$$G_{CM}{}^{1/2}B = 0.8\ B_m. \tag{30}$$

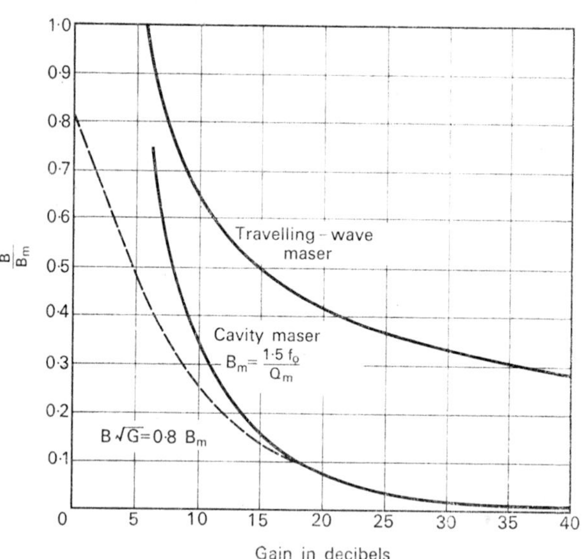

FIG. 1. Normalised maser bandwidth, B/B_m, as a function of gain for the cavity maser and the traveling-wave maser.

The gain–bandwidth curve from (30) is plotted as a dashed line in Fig. 1. It is interesting to note that, for relatively low gains in the cavity maser, the true bandwidth given by (26) is somewhat greater than is estimated from thel imiting gain–bandwidth figure of merit.

The required amplifier slowing and length can be determined from (22). Taking the above example for ruby operation at 6 kmc with Q_m equal to 150 and a structure length of 2 wavelengths (10 cm), we find that a slowing of about 90 is required in the slow-wave structure in order to give a gain of 30 db. An amplifier designed without geometric or resonant slowing, but using the dielectric constant of ruby, would require a length of about 300 cm. The TWM bandwidth for the assumed Lorentzian line shape would be 20 mc. A cavity maser designed using the same material would have a bandwidth of about 1.5 mc. It should also be pointed out that, since a broadband structure is used in the TWM, stagger tuning of the maser material along the maser structure can lead to even greater bandwidths.

The TWM has another important advantage over the cavity maser in that the useful slow-wave structure bandwidth may be an order of magnitude, or more, greater than the maser material bandwidth. Therefore, the center frequency of the maser passband can be tuned electronically over a wide frequency range simply by changing the pump frequency and the dc magnetic field. Thus, a TWM with a 20-mc passband may be tuned over a 200- to 500-mc frequency range at 6 kmc.

An important consideration in maser amplifiers is the sensitivity of the gain to a slight change in the material inversion as measured by χ''. We may, therefore, define the ratio of percentage change in gain to the percentage change in χ'' as a measure of this gain sensitivity, s_g. The gain sensitivity factors for a cavity maser and a traveling-wave maser are respectively,

$$\text{CM}: s_g = \sqrt{G}, \tag{31}$$

$$\text{TWM}: s_g = \log_e G. \tag{32}$$

These two equations are plotted in Fig. 2. They show that, at a gain of 30 db, the stability of a TWM is better by a factor of 4.6

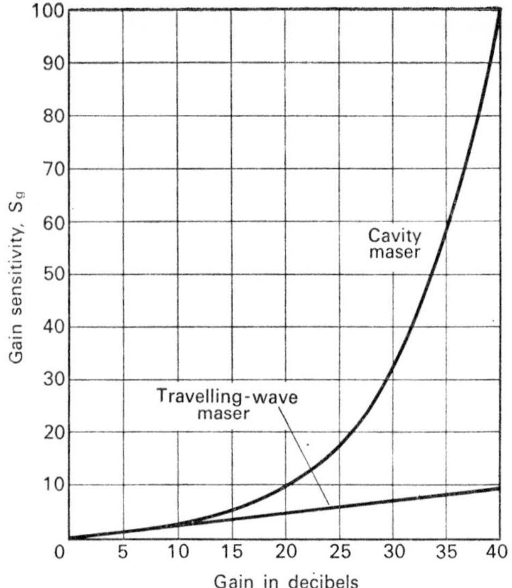

FIG. 2. Pump saturation gain sensitivity, s_g, as a function of gain for the cavity maser and the traveling-wave maser.

than that of the cavity maser. Ultimately, gain stability in a maser is obtained by stabilization of the material χ'' through temperature regulation and regulation of the pump power. It is also advantageous to use sufficient pump power to saturate the pump transition and, hence, make χ'' relatively insensitive to pump power. These techniques are applicable to both the cavity and the traveling-wave maser, but the stability factor, which is given by s_g, is always better in the TWM case. Gain stability may be an important factor in system applications as it has been in noise figure measurements, where the gain fluctuations are equivalent to actual system noise.

The gain stability to load changes of a TWM is also much better than that of a cavity maser. This problem is particularly bad in the case of the cavity maser, because of the dependence of the gain upon the iris coupling factor. This leads to gain changes due to thermal expansion and vibration effects.

The power output of a cavity maser, as well as that of a TWM, is limited by the total volume of active maser material present in the structure. In the case of the cavity maser, this volume is quite small, whereas a TWM uses a long interaction region and the volume of material is greater, often by an order of magnitude. As a result, in the case of the TWM a much wider dynamic range is to be expected, with output powers an order of magnitude greater. It is also possible to optimize output power by proper design of the slow-wave structure in order to increase active material volume.

Just as in the case of the cavity maser, some provision must be made in the propagating structure to allow either a propagating or cavity mode at the pump frequency in order to energize the emissive transition in the material. Finally, ease of fabrication and small size are necessary to the practical realization of the TWM.

III. Slow-Wave Maser Structures

There are three classes of structures suitable for slowing propagation for use with the TWM. The first class uses geometric slowing, such as one obtains in a helix where the energy is propagated on a long circuitous path. The second class uses resonant slowing, as is obtained in a periodic structure in which the energy is internally reflected in the various periods of the structure. The third class employs dielectric slowing and may be used in combination with either of the first two classes. The simple helix structure has an advantage in that it will produce high slowing over a very wide bandwidth of frequency, whereas periodic structures obtain slowing at the expense of tunable bandwidths. However, the circular polarization present on a helix structure has a plane which rotates around the axis of the helix. As a consequence, one requires a spiraling dc magnetic field in order to have the dc field perpendicular to the circularly polarized RF field.

The flattened helix structure of Fig. 3 is a possible broadband slow-wave structure which does have planar regions of circularly polarized magnetic field. The plane of circular polarization is perpendicular to the flat side of the helix and to the direction of propagation. Changes in the sense of polarization are indicated by the arrows.

The second class of slow-wave structures, the periodic type, has a definite passband with associated upper and lower cutoff frequencies. In general, the narrower the structure passband, the

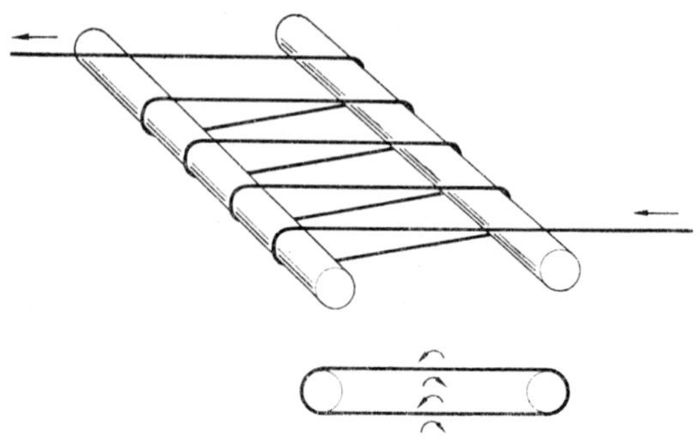

FIG. 3. Flattened helix structure for broadband nonreciprocal gain.

higher is the slowing. It is quite possible to obtain slowing factors of 100 to 1000 in a periodic structure, while 10 to 100 is typical of a helix or other geometrically slowed structure.

It is interesting to note that, while many traveling-wave electron devices require constant phase velocity, the traveling-wave maser requires constant group velocity. As a result, various periodic structures which are very narrowband for tube applications, such as the comb structure, will have wide TWM bandwidths.

Also, it is essential to keep in mind that the structure must propagate at the desired pump frequency. This can be accomplished by propagating the pump power in a waveguide mode and locating the slow-wave structure in the waveguide in such a manner that the two structures are not coupled.

The TE_{10} mode in rectangular waveguide has an equipotential plane, as indicated by the cross section A-A in Fig. 4(a). Thus a planar arrangement of conductors in this cross section will have a minimum of coupling to the waveguide mode. The flattened helix of Fig. 3 could be used in such a waveguide, although some coupling of the two modes would result. Three periodic-type struc-

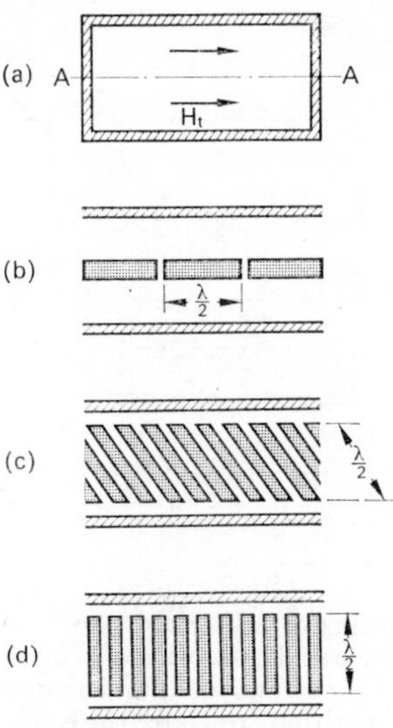

FIG. 4. (a) The equipotential plane in TE_{10} waveguide and three parallel plane structures consisting of half-wave strip-line resonators: (b) end-coupled; (c) slant-coupled; (d) side-coupled.

tures consisting of coupled half-wave strip resonators are shown in Figs. 4(b), (c) and (d). The end-coupled arrangement of Fig. 4(b) is least desirable of these structures because the magnetic field is linearly polarized. The structure of Fig. 4(c), suggested by H. Seidel,[8] has mutual magnetic field coupling, which is variable by changing the slant angle. The structures of both Figs. 4(c) and 4(d) have regions of circular polarization of the magnetic field. The plane of circular polarization is perpendicular to the long dimension of the resonators of Fig. 4(d), and the sense of polarization is opposite above and below the plane of the strips.

Another set of slow-wave structures having planar conductors is shown in Fig. 5. Fig. 5(a) consists of an array of half-wave rods shorted to the side walls. Because of its symmetry, it can be shown that there is no component of Poynting's vector, $E \times H$, in the

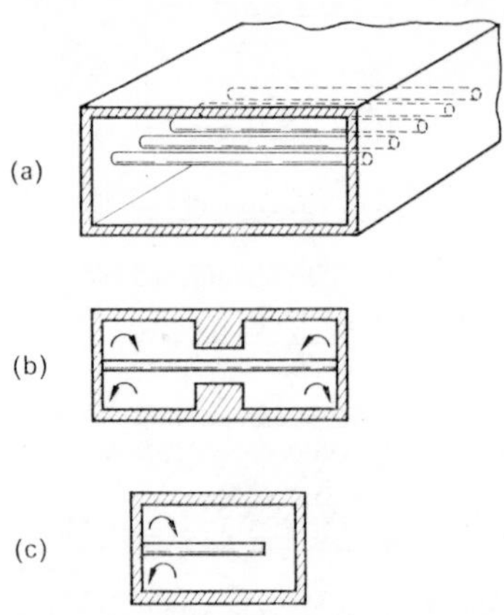

FIG. 5. Structures consisting of planar arrays of conductors with electrical connection to the waveguide walls; the plane of circular polarization is perpendicular to the fingers and the arrows indicate changes in the sense of polarization: (a) Easitron zero passband structure; (b) Karp propagating structure; (c) comb propagating structure.

direction of the waveguide propagation. Consequently, it is a non-propagating structure and has been suggested for use in the Easitron.(9) It is interesting because it points out the separate effects of electric and magnetic coupling between adjacent rods. A perturbation of the enclosing waveguide, such as that of the Karp(10) type of structure, in Fig. 5(b), produces a propagating passband in the structure. Thus, the Easitron structure may be called a zero passband structure in which the effect of electric and magnetic coupling between rods just cancels. The general properties of propagating parallel arrays has been discussed by Pierce.(11)

Although the structure of Fig. 4(b) is similar to the Easitron, this structure will propagate, due to the fringe capacity at the ends of the resonators. The comb structure of Fig. 5(c) will have characteristics very similar to Fig. 4(d), since it is essentially Fig. 4(d) with a shorting plane along the center line, which is an equipotential plane for the slow-wave structure. A comb-type structure was used by Millman.[12] This structure had rather broad fingers, however, as required for effective electron beam interaction.

A rather extensive study of periodic slow-wave structures is given by Leblond and Mourier.[13, 14] A method for calculating the characteristics of such structures which takes into account coupling between nonadjacent wires was developed by Fletcher[15] and applied to the interdigital line. The interdigital line may be thought of as two comb structures attached to opposite walls of the waveguide with fingers interleaved. The interdigital line does not, in general, produce as much slowing as a comb structure, since, in its passband, the signal wave is propagated at the velocity of light along the circuitous path between adjacent fingers. Thus, the propagation does not depend upon a critical balance of finger to finger coupling and can take place without fringe capacity at the finger tips or a side-wall perturbation. A number of design curves for digital structures have been computed by Walling[16] using the theory of Fletcher. Nonreciprocal attenuation in the interdigital circuit has been analyzed and measured by Haas.[17]

IV. The Comb-Type Slow-Wave Structure

As indicated by the arrows in Fig. 5(c), the comb-type slow-wave structure has regions of circularly polarized magnetic field above and below the plane of the fingers. Since the sense of polarization is reversed in the two regions, the structure is particularly suited to the TWM application.[18] The maser material can be placed on one side of the fingers and the isolator material on the opposite side. The magnetic field varies from a maximum at the shorted end of the rod to zero at the other end. Since the electric field does just the opposite, it is possible to place a high dielectric material, such as ruby, in a region where effective magnetic interaction is obtained without incurring substantial dielectric loading, which might adversely affect the structure characteristics.

The microwave pump power is propagated through a waveguide enclosing the comb structure, such as in Fig. 6. The TE_{10} waveguide mode will produce strong longitudinal fields near the wave-

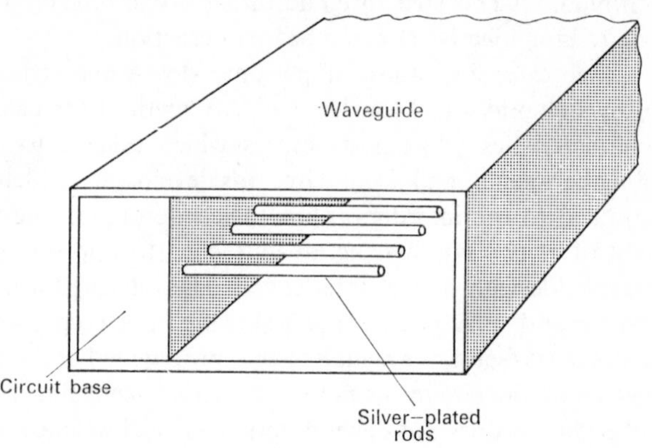

FIG. 6. The comb-type slow-wave structure.

guide wall and transverse fields in the center of the guide. The dc magnetic field is applied in the direction of the fingers of the structure. Thus, for a maser crystal against the waveguide wall nearest the base of the fingers, $\Delta S = \pm 1$ pump transitions would be excited, while $\Delta S = 0$ pump transitions would be most strongly excited for a centrally located crystal. In our operation, both gadolinium ethyl sulfate and ruby maser materials have $\Delta S = 0$ pump transitions. Thus, it is necessary either to move the crystal away from the waveguide wall slightly, which reduces the gain, or to use a higher pump power, in order to invert the spin population throughout the crystal. The waveguide structure can, of course, be extended beyond the slow-wave structure. A waveguide short and coupling iris can then be used to obtain resonant enhancement of the pumping fields and, hence, better coupling to the pump transition in the crystal.

A perturbation measurement has been made, using a small sphere of ferrimagnetic material to measure the circularly polarized components of the magnetic field in such a comb structure. The absorption of the sphere at ferrimagnetic resonance is a measure

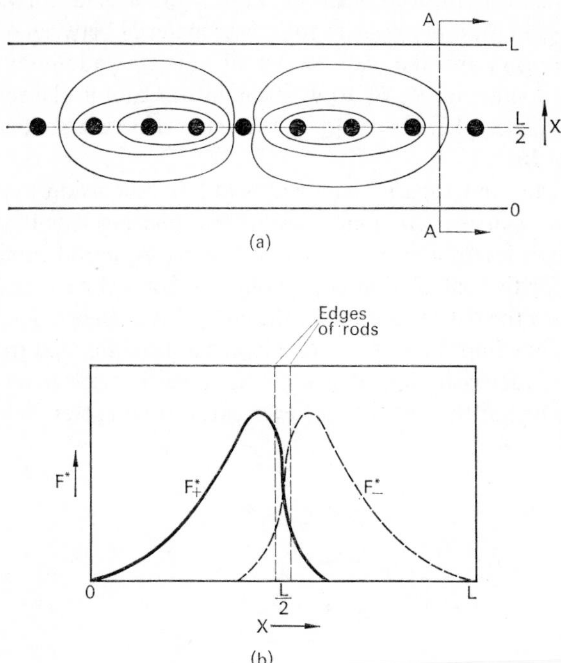

FIG. 7. The magnetic field pattern (a) and measured amplitudes of circular polarized field components (b) for the comb-type slow-wave structure.

of the circularly polarized filling factor per unit volume, which we will call $F_+{}^*$. The result of such a measurement is plotted in Fig. 7. The dashed line in Fig. 7(b), labeled $F_-{}^*$, is obtained either by reversing the direction of propagation through the structure or by reversing the dc magnetic field. The circularly polarized filling factor, F_+, and the figure of merit, R, can be obtained by integrating the appropriate curve over the cross section. This one measurement is not sufficient to get the exact filling factor because the structure is not uniform. The filling factor should be calculated from measurements made over the volume of one period of the structure. The measurement of Fig. 7 predicts a forward-to-reverse gain ratio of about 15 for maser material between one side of the fingers and the wall. In actual tests on gadolinium ethyl sulfate, a value of R_m of 10 was obtained. Careful placement of ferrimagnetic sphere isolators in the same structure gave a value for R_i of 30.

The gain and tunable bandwidth of a maser using the comb structure can be computed from the phase-versus-frequency characteristics of the slow-wave structure. A useful equivalent circuit for this calculation is given in Fig. 8(a). The capacity, C_1, represents the fringe capacity at the ends of the fingers. The transmission line impedances, Z_{01} and Z_{02}, can be computed from the TEM characteristic impedance of the finger in waveguide transmission line at the upper and lower cutoff frequencies. When the

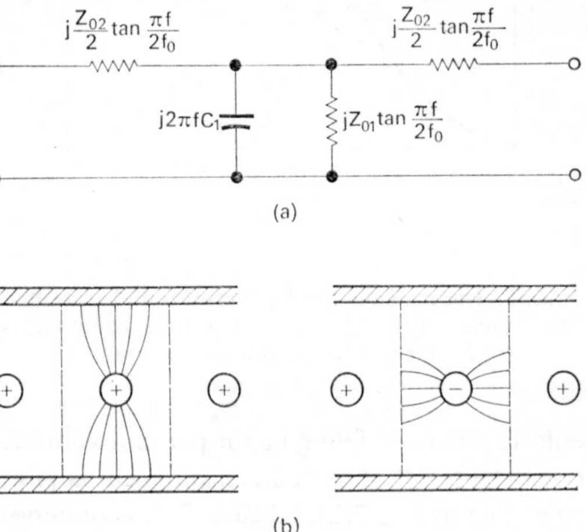

FIG. 8. (a) Equivalent circuit for the comb structure and (b) the cutoff-frequency TEM mode electric field patterns.

phase shift between fingers is zero, the electric field pattern, Z_{++} is that shown on the left in Fig. 8(b). The electric field pattern at the other cutoff frequency of the structure, Z_{+-}, is that corresponding to π phase shift, indicated on the right in Fig. 8(b). The impedances Z_{++} and Z_{+-} can be obtained analytically or by means of the usual electrolytic tank or resistance card analog computers. These cutoff impedances are related to those in the equivalent circuit by

$$Z_{01} = Z_{++}, \tag{33}$$

$$Z_{02} = \frac{4Z_{+-}}{1-\dfrac{Z_{+-}}{Z_{++}}}. \tag{34}$$

The determinantal equation for the circuit of Fig. 8(a) is

$$2\pi f C_1 + \left(\frac{1}{Z_{01}} + \frac{4}{Z_{02}} \sin^2 \varphi/2\right) \tan \frac{\pi f}{2f_0} = 0, \tag{35}$$

where φ is the phase shift per section and f_0 is the frequency at which the circuit fingers have an electrical length of one-quarter wavelength. Solution of this transcendental equation is simplified by use of the curve in Fig. 9. The abscissa is the signal frequency normalized to f_0. The ordinate is $X_{co}Y(\varphi)$, which is given by

$$X_{co} = \frac{1}{2\pi f_0 C_1}, \tag{36}$$

$$Y(\varphi) = \frac{1}{Z_{01}} + \frac{4}{Z_{02}} \sin^2 \varphi/2. \tag{37}$$

The signal frequency for a certain phase shift is obtained by computing $X_{co}Y(\varphi)$, and then referring to Fig. 9.

The product SN required in the gain calculation can be obtained from experimental measurement of φ versus frequency from the equation

$$SN = N_s \frac{f_0}{2\pi} \frac{d\varphi}{df}, \tag{38}$$

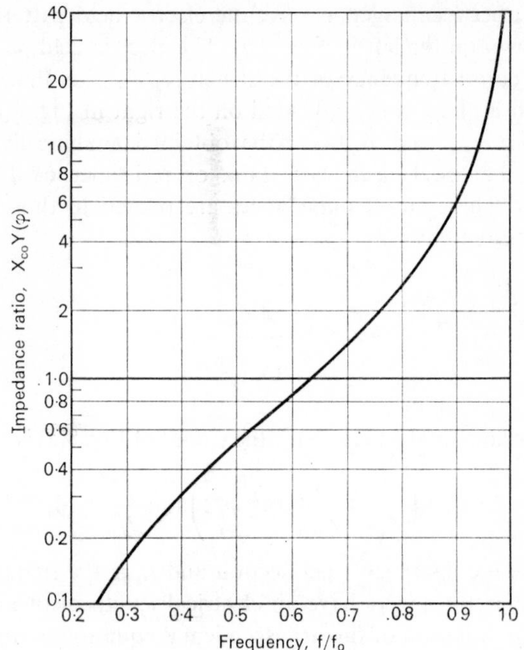

FIG. 9. The curve, X_{co} $Y(\varphi)$ vs. f, used to solve equation (35).

where N_s is the total number of sections in the structure. For the equivalent circuit of Fig. 8(a), the SN product can be obtained from the curve of Fig. 10, which gives the quantity s, defined by

$$s \equiv \frac{SNX_{co}}{N_s Z_{02}} \sin \varphi. \tag{39}$$

Experimentally measured values of φ versus f were compared with the values calculated from the equivalent circuit, for a comb structure in which the end capacity, C_1, could be varied. The resulting calculations are shown in Fig. 11.

The three curves were computed from the equivalent circuit analysis using the measured values of Z_{++} and Z_{+-}. The fringe

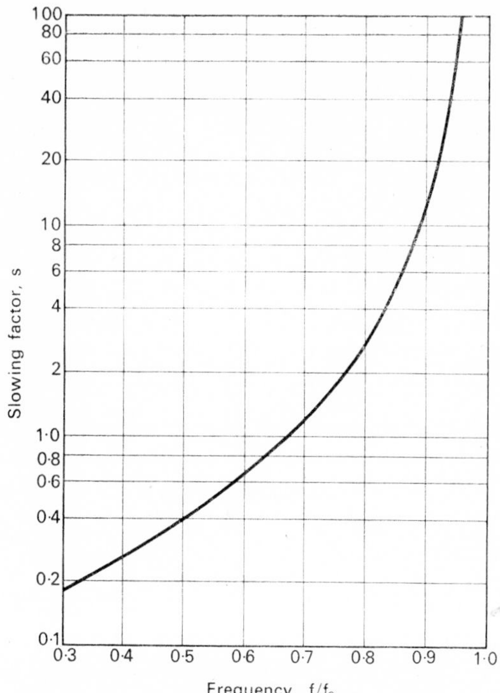

FIG. 10. The slowing factor, s, as a function of frequency, used to determine SN from (39).

capacity was determined by using the measured frequency for 0.2π phase shift. It should be mentioned that the equivalent circuit does not fully take into account the effect of coupling between nonadjacent fingers, an effect which may be important in this particular circuit.

A rather wide passband was obtained with the particular choice of parameters, even with rather large spacing between the finger ends and the opposite waveguide wall. Also, because of the wide bandwidth, the slowing varies over a wide range. The increase in slowing at the high-frequency end of the band is partially com-

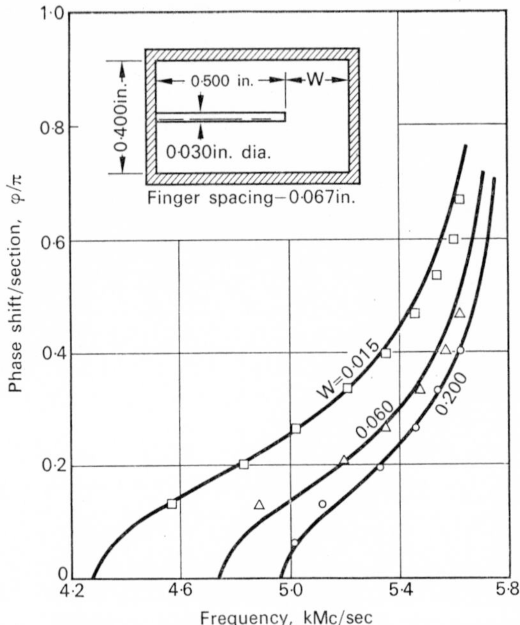

FIG. 11. Phase shift per section, φ, vs. frequency, showing the effect of end wall capacity upon slowing.

pensated for by the reduction in the filling factor due to an increase in the stored energy between the fingers.

The structure passband can be decreased by reducing the waveguide height and increasing the distance between fingers. Too great a reduction in waveguide height will make the comb fingers a rather serious perturbation of the waveguide pump mode and will degrade the maser gain nonreciprocity factor, R_m. On the other hand, increasing the distance between fingers increases the over-all length of the amplifier.

Several methods for reducing the bandwidth of the comb structure have been tried. The first consisted of building a structure in

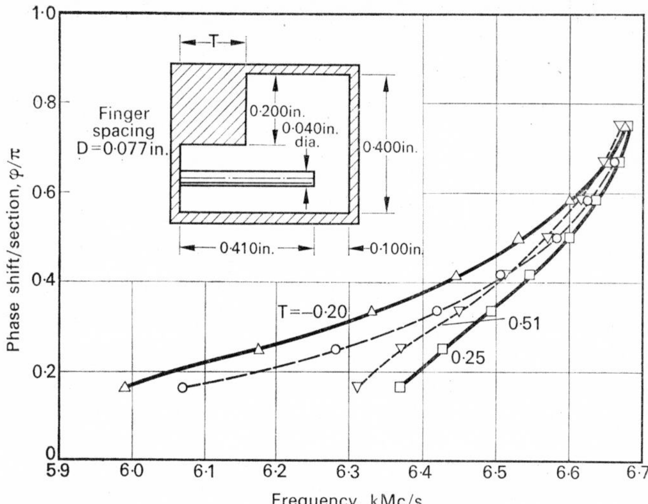

FIG. 12. Phase shift per section, φ, vs. frequency showing the effect of a side wall perturbation upon slowing.

which a wall perturbation could be moved across the width of the structure. The cross section of this test structure is shown in Fig. 12, along with the measured results obtained. The curve for $T = -0.2$ inch is for the case where the sliding side wall perturbation is pulled back from the base of the fingers. In this case, the bandwidth is increased. However, with $T = 0.25$ the bandwidth is reduced by a factor of 2 and relatively constant slowing is obtained over the band.

A second method for reducing the bandwidth is that shown in Fig. 13. The tips of the structure fingers are dielectrically loaded with a polystyrene strip having holes for the fingers. The dielectric increases the finger-to-finger capacity without greatly increasing the finger-to-wall capacity. Thus, this capacity would be in shunt with Z_{02} in the equivalent circuit of Fig. 8. It is apparent from the results in Fig. 13 that very narrow bandwidths can be obtained in this manner. However, the variations in structure dimensions

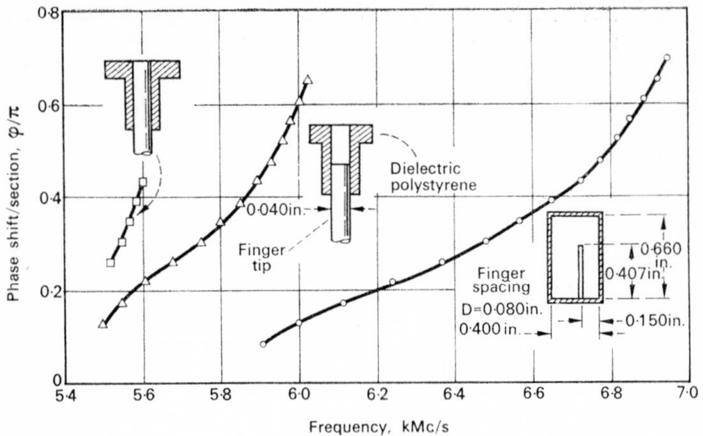

FIG. 13. Phase shift per section, φ, vs. frequency, showing the effect of finger-to-finger capacity upon the slowing.

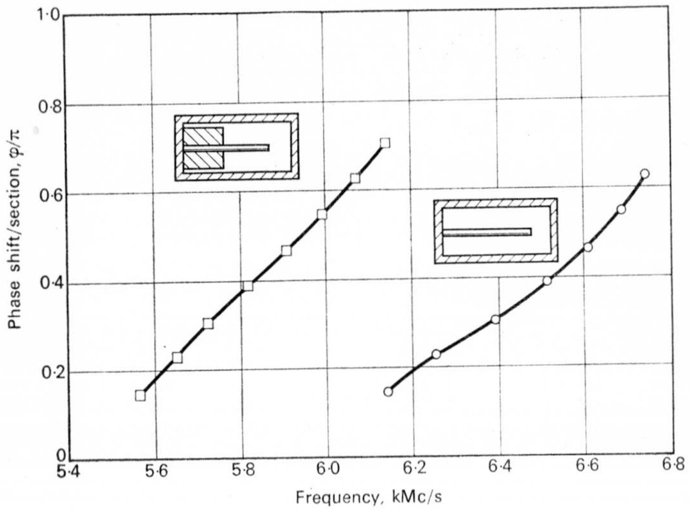

FIG. 14. Phase shift per section, φ, vs. frequency for the comb structure used in the full-length ruby maser; the unloaded structure characteristic is compared with that of the ruby loaded structure.

become rather critical. Thus only a few points on the φ–f curve could be obtained using the structure resonance. This was also due in part to the high losses resulting from the high slowing.

The full-length ruby maser employed a slow-wave structure whose φ–f characteristics are shown in Fig. 14. Notice that the unloaded structure has a wider passband than the final amplifier with dielectric loading of the ruby on both sides of the structure. In this case, the slowing was improved by not loading the ruby to the full height of each side of the waveguide. Thus, in this case, bandwidth narrowing is obtained by selective location of the maser material itself.

A number of different coupling schemes have been employed to match a 50-ohm coaxial cable into the comb structure. The matching arrangement shown in Fig. 15 gives quite broadband results. It has been found that a good impedance match can be obtained only over that frequency range for which the slowing factor is relatively constant. A VSWR less than 1.5 is typical over the useful band of the structure. Rapid variation of measured VSWR is observed which can be attributed to periodic variations in the structure which result in internal resonances. In actual

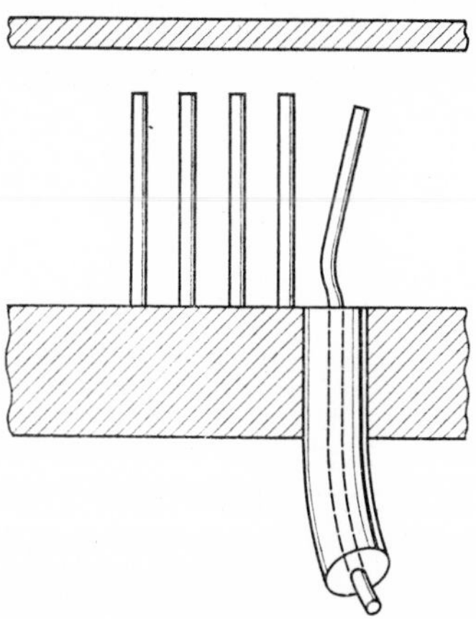

FIG. 15. Detail of the coaxial-to-comb-structure impedance match.

maser operation, these internal resonances should be suppressed by the nonreciprocity of the structure.

V. Gadolinium Maser Test Section Results

The first TWM tests were performed using gadolinium ethyl sulfate as the active maser material and yttrium iron garnet as the isolator material. The cross section of this TWM is shown in Fig. 16. The slow-wave structure was similar to that of Fig. 12, since

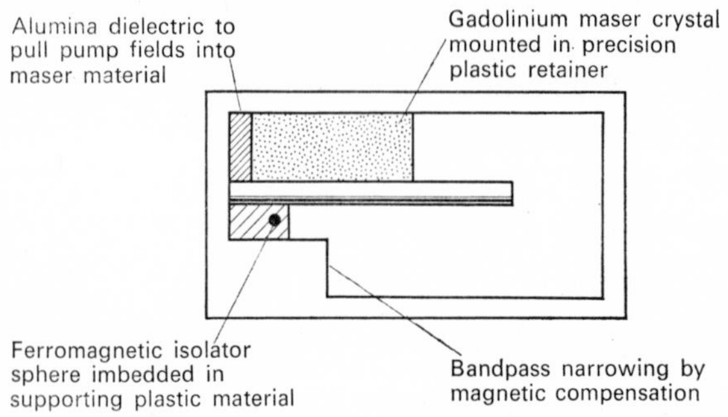

FIG. 16. Cross section of gadolinium maser, showing location of active material and polycrystalline yttrium iron garnet isolators.

it employed a perturbation of the side wall to narrow the passband. Of course, the maser material dielectric constant also affected the structure characteristics and a certain optimum loading could be obtained.

Since the maser operation in gadolinium requires the pump magnetic field to be parallel to the applied magnetic field, a slab of dielectric was added against the waveguide wall to enhance the transverse RF waveguide magnetic field.

The "impurity dope" maser action in cerium-doped gadolinium ethyl sulfate produced at 1.6°K a magnetic Q_m of 170 in the structure used. The signal frequency was 6.0 to 6.3 kmc, and the pump frequency was 11.7 to 12.3 kmc. The magnetic field was about

1800 gauss. The nonreciprocity factor, R_m, was measured as about 10. The maser gain obtained was 12 db with about 1 inch of the slow-wave structure filled with the gadolinium salt. The active material consisted of three separate crystals which had to be accurately cut and aligned. Thus, it became apparent that the physical properties of the gadolinium salt were not too well suited for more than laboratory tests. The gadolinium bandwidth, B_m, was 30 mc.

It should be mentioned, however, that the maser operation in gadolinium does have the advantage of high saturation power and fast saturation recovery time. A power output of +15 dbm was obtained and the saturation recovery time was measured as about 20 microseconds.

The magnetic field required by the maser material is in the vicinity of that for ferrimagnetic resonance in a sphere. Hence, it was possible to use spheres of a gallium-substituted yttrium iron garnet, which was found to have a relatively high χ'' at liquid helium temperature. Spheres of yttrium iron garnet were located in the A-A cross section indicated in Fig. 7(a). By careful location of the spheres, approximately one per finger, an isolation ratio of 50 was obtained at room temperature. The isolation ratio at 1.5°K was 30.

VI. Ruby Traveling-Wave Maser Performance

The first high-gain full-length TWM was constructed using pink ruby which had approximately 0.05 per cent Cr^{+++} in the Al_2O_3 parent crystal. The slow-wave structure is shown in Fig. 17 with one side of the waveguide removed. Two pieces of ruby are shown in position and a third is in the foreground. The comb structure has a length of 5 inches and consists of 62 brass rods approximately 0.4 inch long. The phase-shift characteristics of this structure were given in Fig. 14.

The holes for insertion of the coaxial input and output matches are visible at the ends of comb structure of Fig. 17. Fig. 18 is a cutaway drawing of the complete TWM assembly, showing the

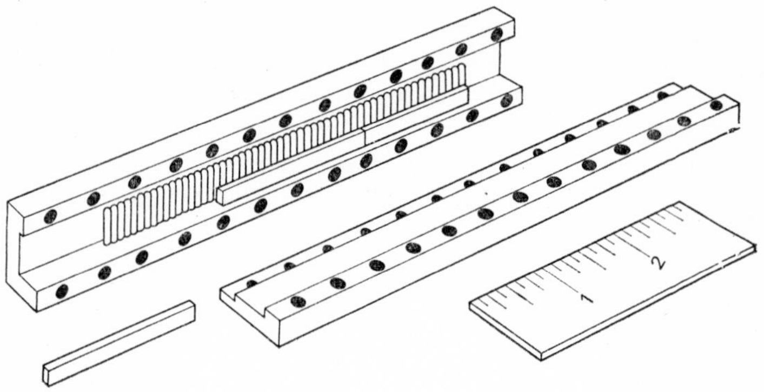

FIG. 17. View of the disassembled comb structure for the ruby maser, with two pieces of ruby material shown in position.

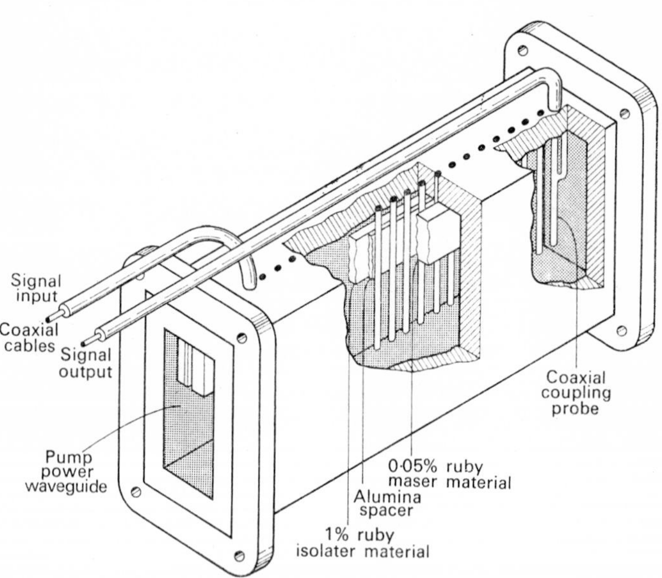

FIG. 18. Cutaway view of full-length ruby maser assembly.

waveguide flanges which are added to the ends of the structure for attachment of a movable short and the pump waveguide coupling iris. This drawing also shows the location of the 0.05 per cent ruby amplifying material and the 1 per cent ruby isolator material. Notice that the 1 per cent ruby is spaced away from the circuit fingers by an alumina slab. This reduces the coupling of the isolator to the forward propagating circuit wave, thus improving the isolation ratio, R_i.

This TWM gives a net forward gain of 23 db and a net reverse loss of 29 db, including the input and output coaxial cables extending into the 1.5°K helium bath. With the dc magnetic field off, the structure gave a loss, L_0, of 3 db. The electronic gain of the maser material, the first term in (19), was 30 db, and the loss of the isolator was 35 db. Thus, the ratios R_m and R_i are about 3.5 and 8.5 respectively. The improved figure of merit for the isolator is due to the addition of the alumina slab. The high loss of the isolator, in spite of the spacer, is due to the increased Cr concentration.

The 3-db bandwidth of the TWM was measured as 25 mc at a center frequency of 5.8 kmc. This bandwidth is somewhat in excess of that predicted by an assumption of Lorentzian line shape as used in Fig. 1. The passband of the structure allows amplification over a frequency range of 5.75 to 6.1 kmc. The pump frequency must be tuned from 18.9 to 19.5 kmc and the dc magnetic field varied from 3.93 to 4.07 kilogauss to cover this electronic tuning range.

At a power output of -22 dbm, the gain of the amplifier is reduced by 0.5 db. The recovery time of the amplifier after saturation by a large signal is quite long, being on the order of 10^{-1} sec. This is due to the long relaxation time in ruby. It has one compensation, however, in that spin storage is very effective for increasing the pulse saturation power. Thus, with a pulse length of 10 microseconds and a pulse repetition rate of 100 per second, a pulse power output of $+8$ dbm was measured for the same 0.5-db reduction in gain. The saturation characteristic of a TWM is not abrupt, but is a smooth curve going to lower gain as the power is increased. In the limit of very high power, the TWM is essentially transparent. It will have an insertion loss of about 3 db, since after the maser material saturates, it produces neither gain nor loss.

The pump source had a power output of 100 mw. Recent experiments have been performed which show that satisfactory operation of the ruby TWM can be obtained with no waveguide iris and a short at the end of the amplifier. In this type of operation the pump power absorbed by the amplifier was less than 10 mw. The pump power absorbed by the amplifier determines the refrigeration power input required in a continuously cooled system. Thus a factor-of-10 reduction in maser pump power has a substantial effect upon the size of refrigerator required.

As was mentioned previously, it should be possible to obtain greater bandwidths in a TWM by stagger-tuning of the maser material. This was verified in this amplifier by rotating the maser magnet, which resulted in tuning of the three maser crystals to different frequencies. A bandwidth of 67 mc at a gain of 13 db was measured.

VII. TWM Noise-Temperature Measurement

Because of the high gain stability inherent in the TWM it was possible to perform a quite accurate measurement of noise temperature. Also, since the TWM is a unilateral two-port amplifier by itself, no external isolators or circulators are required when it is used, for instance, as a radar preamplifier.

The experimental system for the noise-temperature measurement is shown in Fig. 19. Two noise sources, which are matched loads maintained at different temperatures, are connected alternatively to the TWM input with an electrically operated waveguide switch. Isolators were included at the input and output of the TWM. Although the TWM is short-circuit stable, small gain fluctuations can be produced by changes in the input VSWR. Since gain fluctuations on the order of 0.02 db are significant in this measurement, an input isolator was included to eliminate gain changes when the waveguide switch is operated. An isolator

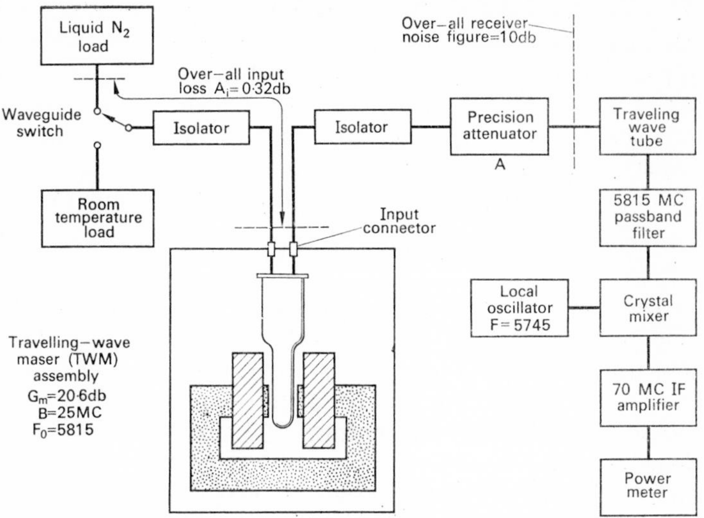

FIG. 19. Experimental system for TWM noise temperature measurement.

was included on the output of the TWM to insure that no excess noise from the traveling-wave tube (TWT) would be fed back into the TWM and prevent maser gain changes because of the precision attenuator setting.

The microwave detecting system consisting of the TWT, filter, crystal mixer, etc., had an over-all noise figure of 10 db. The actual noise temperature measurement is made by switching the noise source and adjusting the precision attenuator to maintain constant output power at the power meter.

A total of 37 separate measurements were made over a period of about 30 minutes, and the resulting noise temperature was calculated to be

$$T_m = 10.7 \pm 2.3°\text{K}.$$

Approximately two-thirds of the estimated error was a statistical variation in the observed attenuator reading. About half of this attenuator reading variation can be attributed to gain fluctuations in the system of about ± 0.03 db during any one measurement and the other half to observational error. The remaining error is principally due to an error of ± 0.02 db in the determination of the input circuit loss, A_i.

Since the above-measured noise temperature is referred to an input connector at room temperature, it is a useful system temperature and the noise figure of the resulting preamplifier TWM is

$$F = 1.037 \pm 0.008 \ (0.16 \pm 0.03 \text{ db}).$$

The TWM employs half-inch diameter 50-ohm coaxial input and output leads of low heat conductivity monel to reduce heat loss from the liquid helium bath which was maintained at a temperature of 1.6°K. The coaxial cables were silver plated and polished to reduce microwave losses. The input cable, 30 inches long, had a room temperature loss of 0.28 ± 0.02 db. The actual input cable temperature gradient was monitored with 12 thermometers throughout its length. A calculation, using this data and taking into account the known variation of the resistivity of silver with temperature, shows that a noise temperature of 9 ± 1°K is produced by the input cable losses.

Another possible type of noise is that fed back into the TWM output from the isolator at room temperature. In the TWM, this produces a negligible contribution at the input because of the reverse isolation.

The noise contributed by the maser proper depends upon the spin temperature and the ratio of spin system gain to the circuit loss. A theoretical noise temperature of 2.4 ± 0.2°K is thus calculated from maser noise theory.[19, 20, 21]

The theoretically calculated over-all TWM noise temperature is then

$$T_m(\text{theory}) = 11.4 \pm 1.2 \text{°K}.$$

Since it is apparent that most of the above noise is contributed by the input cable, a second experiment is planned which will exclude the input cable loss and allow a more direct measurement of the actual noise temperature of the maser amplifier proper.

VIII. Future Studies

An extensive investigation of the ferrimagnetic properties of various ferrite and garnet materials at liquid helium temperatures is being carried out by F. W. Ostermayer of Bell Telephone Laboratories. A number of materials look promising for application as isolators in the ruby TWM. A ferrimagnetic isolator can be expected to have lower forward loss, higher reverse isolation, and low pump-power absorption.

Further reduction of the TWM tunable bandwidth will increase the gain. Thus it should be possible to obtain useful wideband gain at a somewhat higher bath temperature, such as 4.2°K.

It appears that, by using high pump-frequency-to-signal-frequency ratios, it will be possible to build low-frequency traveling-wave masers with about the same percentage tunable bandwidth as the present 6 kmc amplifier. Accordingly, a number of suitable circuits are being investigated.

IX. Acknowledgment

The initial slow-wave structure investigation for TWM application was begun by H. Seidel, whose assistance is gratefully acknowledged. We would also like to thank A. Pohly for technical assistance and G. Shimp for computation and plotting of curves.

REFERENCES

(1) BLOEMBERGEN, N., Proposal for a new-type solid state maser, *Phys. Rev.* **104,** October 1956, p. 324.

(2) SCOVIL, H. E. D., FEHER, G. and SEIDEL, H., Operation of a solid state maser, *Phys. Rev.* **105,** January 1957, p. 762.

(3) SCOVIL, H. E. D., The three-level solid state maser, *Trans. I.R.E.*, **MTT-6,** January 1958, p. 29.

(4) MCWHORTER, A. L. and MEYER, J. W., Solid state maser amplifier, *Phys. Rev.* **109,** January 15, 1958, p. 312.

(5) MAKHOV, G., KIKUCHI, C., LAMBE, J. and TERHUNE, R. W., Maser action in ruby, *Phys. Rev.* **109,** February 15, 1958, p. 1399.

(6) SCHULZ-DU BOIS, E. O., Paramagnetic spectra of substituted sapphires—Part I: ruby, *B.S.T.J.*, **38,** January 1959, p. 271.

(7) SCHULZ-DU BOIS, E. O., SCOVIL, H. E. D. and DE GRASSE, R. W., *Bell Syst. Tech. J.* **38**, March 1959, p. 335.
(8) SEIDEL, H., private communication.
(9) WALKER, L. R., unpublished manuscript.
(10) KARP, A., Traveling-wave tube experiments at millimeter wavelengths with a new, easily built space harmonic circuit, *Proc. I.R.E.* **43**, January 1955, p. 41.
(11) PIERCE, J. R., Propagation in linear arrays of parallel wires, *Trans. I.R.E.* **ED-2**, January 1955.
(12) MILLMAN, S., A spatial harmonic amplifier for 6-mm wavelength, *Proc. I.R.E.* **39**, September 1951, p. 1035.
(13) LEBLOND and MOURIER, Étude des lignes à barraux à structure périodique pour tubes électroniques U.H.F., *Ann. de Radioélect.* **9**, April 1954, p. 180.
(14) LEBLOND and MOURIER, Étude des lignes à barraux à structure périodique —deuxième partie, *Ann. de Radioélect.* **9**, October 1954, p. 311.
(15) FLETCHER, R. C., A broadband interdigital circuit for use in traveling-wave-type amplifiers, *Proc. I.R.E.* **40**, August 1952, p. 951.
(16) WALLING, J. C., Interdigital and other slow-wave structures, *J. Elect. and Cont.* **3**, September 1957, p. 239.
(17) HAAS, L. K. S., *Unilateral Attenuation in the Interdigital Circuit*, WADC TR 57–239, Wright-Patterson Air Force Base, Ohio, May 1957.
(18) DEGRASSE, R. W., *Slow-Wave Structures for Unilateral Solid State Maser Amplifiers*, I.R.E., WESCON Conv. Record, August 1958.
(19) SHIMODA, K., TAKAHASI, H. and TOWNES, C. H., *J. Phys. Soc. Japan*, **12**, 1957, p. 686.
(20) POUND, R. V., *Ann. Phys.* **1**, 1957, p. 24.
(21) STRANDBERG, M. W. P., Inherent noise of quantum-mechanical amplifiers, *Phys. Rev.* **106**, May 15, 1957, p. 617.

PAPER 16*

Solid State Masers and Their Use in Satellite Communication Systems

J. C. WALLING † and F. W. SMITH †

Summary

The design of a maser for practical applications calls for a combination of electronics and solid state physics. The article below discusses this combination. A full description is also given of the travelling wave maser which was built at the Mullard Research Laboratories for use in experiments with the satellites Telstar and Relay.

Introduction

It is now well known that the solid state maser amplifier provides the most sensitive available means of amplifying microwave radiation. It is potentially capable of permitting the detection of signals in the microwave region having energies of only a few quanta. The device has therefore excited considerable interest among radio astronomers and more recently among communications engineers. Indeed the spectacular success of the recent satellite communication projects Telstar and Relay was due, in no small part, to the use of solid state maser amplifiers in the receiving stations.

It is the purpose of this article to discuss the design, performance and applications of travelling wave maser (TWM) amplifiers with particular reference to the maser recently developed at Mullard Research Laboratories for use at the Communication Satellite Earth Station of the British General Post Office at Goonhilly

* *Philips Tech. Rev.* **25,** 289–90, 304–8 (1963/4). (Reproduced only in part).

† Mullard Research Laboratories, Redhill (Surrey), England.

Down, Cornwall. A preliminary account of this device has already been given elsewhere.[(1)*]

Satellite Communications

The first experiments in communication via an artificial earth satellite were carried out in 1960 with equipment designed and constructed by Bell Telephone Laboratories and using the 30 m diameter metallized balloon Echo I[(2)] which was launched from Cape Kennedy into an almost circular orbit in August 1960 at an altitude of about 1600 km. Echo acts as a passive reflector of signals from a ground transmitter, and the scattered signal is received at other ground stations.

The path loss over the complete circuit may be shown to be

$$L = \frac{(4\pi)^3\, d_1{}^2 d_2{}^2}{G_1 G_2\, \lambda^2 \sigma},$$

where d_1 and d_2 are the path lengths from the two ground stations to the satellite, G_1 and G_2 are the aerial gains over an isotropic radiator appropriate to the two stations, λ is the wavelength and σ is the scattering cross section of the satellite; see Fig. 1.

From this expression the magnitude of the received signal can be calculated. Assuming that the transmitter and receiver aerials have gains of about 45 dB (a practical figure), then with the satellite (scattering cross section $\sim$730 m^2) in the most favourable position with respect to two stations 4800 km apart, we find that the path loss at 2400 Mc/s is about 180 dB and therefore with a 10 kW transmitter the received carrier power is 10^{-14} watt; signals of this order may easily be swamped by noise. For example, a good low noise receiver with 1 Mc/s bandwidth and a noise factor

* *Note of the editor:* In view of the fact that the Gaussian system is in common use in the current literature on masers, we have refrained from converting to the rationalized Giorgi system.

(1) J. C. Walling and F. W. Smith, *Brit. Commun. and Elec.*, Aug. 1962.

(2) *Bell Syst. Tech. J.* **40,** No. 4 (1961).

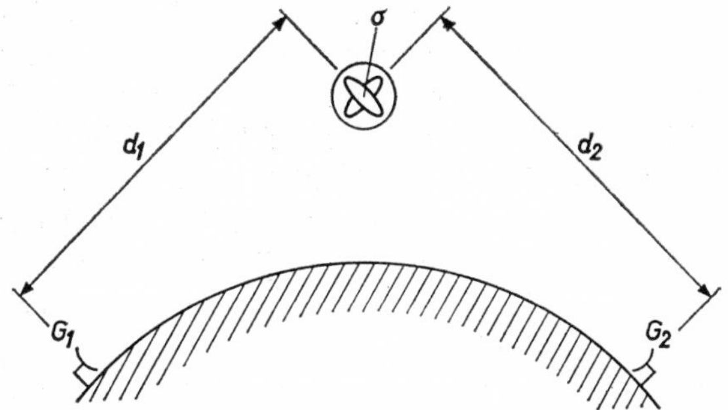

FIG. 1. Satellite communication scheme. The satellite has a scattering cross section σ, and the ground station aerials have gains of G_1 and G_2.

of 1.06 dB (equivalent noise temperature[3] of 50°K) has an equivalent input noise power of 0.7×10^{-15} watt and could thus, for this case, only realize a carrier-to-noise power ratio of about 10 dB.

Initial experiments were carried out with receivers of much narrower bandwidth than this to achieve satisfactory signal-to-noise ratios, and although the experiments were successful the use of a narrow bandwidth always severely limits the amount of information which can be put over the channel.

In a world-wide communication system it is desirable to use satellites at greater altitudes than the 1600 km at which Echo I is

(3) When we speak of the equivalent noise temperature T of an amplifier we mean that the noise added to the signal by the amplifier is equivalent to that which would be generated by a matched resistive termination at temperature T radiating into the input waveguide of the amplifier. However the noise factor F of the amplifier is, according to the standard definition:

$$F = \frac{\text{input signal}}{\text{input noise}} \Big/ \frac{\text{output signal}}{\text{output noise}}$$

when the input is generated by a matched source at a temperature $T_s = 290$°K. Hence:

$$F = 1 + \frac{T}{290}.$$

orbiting because of the need to obtain longer periods of mutual visibility between ground stations. However this invariably implies an even greater path attenuation and even smaller received signals.

The second generation of communication satellites, Telstar, developed by the Bell Laboratories and launched in 1963[(4)] and Relay, developed by R.C.A. for N.A.S.A. and launched in 1963,[(5)] employs active repeaters in the satellite. The signal from one ground station is received in the satellite, frequency converted, amplified and re-radiated at a power level of about 2 watts (Telstar I and II) and 10 watts (Relay I). By introducing gain in the system in this way it is possible to work at greater satellite altitudes with longer periods of mutual visibility between ground stations and also to increase the bandwidth of the system. Furthermore, use of these greater altitudes is desirable in order that the active satellites avoid the van Allen radiation belts. More sophisticated aerials with gains above 55 dB have been built for these experiments and signals received from Telstar I at slant ranges (e.g. d_1 in Fig. 1) of between 4000 and 10 000 kilometres were in the range between 10^{-12} and 10^{-13} watt. This follows from the ratio of power received at the aerial P_A to that radiated substantially isotropically by the satellite P_S which is:

$$\frac{P_A}{P_S} = \frac{G\lambda^2}{16\pi^2 d^2} \text{ (}G\text{ being the aerial gain).}$$

The bandwidth of the system using an active repeater in the satellite may be increased to about 25 Mc/s since the noise power introduced by a receiver with such a bandwidth and an equivalent noise temperature of 50°K is now 1.75×10^{-14} watt and even at ranges up to 10 000 kilometres the carrier-to-noise power ratio is greater than 10 dB. Sophisticated modulation techniques such as wideband frequency modulation with FM feedback in the receiver are currently used to improve further the overall signal-to-noise ratio.[(5)]

(4) "The Telstar experiment", *Bell Syst. Tech. J.* **42,** No. 4 (1963).

(5) *Programme and Conference Digest*, I.E.E. Conference on Satellite Communications, Session 8, Nov. 1962.

From the foregoing it is apparent that receivers with noise temperatures of the order of 50°K or less are necessary if wide band satellite communication systems are to be successfully realized with practical satellite transmitter powers which, in the present state of the art, are limited to a few watts. Clearly then it is necessary to pay careful attention to the reduction of all sources of noise in the receiving system and it is thus desirable that the first stage amplifier should have the lowest possible noise temperature; this amplifier should then be a maser.

A 4170 Mc/s Packaged Travelling Wave Maser

As an example of a practical packaged TWM we will now describe briefly the amplifier designed and built at the Mullard Research Laboratories and used at the General Post Office satellite communication ground station at Goonhilly Down in Cornwall for experiments with the Telstar and Relay communication satellites.

Maser Structure

The active material in this maser is 0.05% ruby used with the applied static magnetic field normal to the three-fold axis of symmetry of the crystal as described in the last section. The signal transition is that between levels 1 and 2 at 4170 Mc/s and a population inversion of about 2.7 is obtained by pumping between levels 1 and 4 at 30 150 Mc/s. These frequencies correspond to a magnetic field of 3280 Oe.

The ruby sample was cut from a large synthetic single crystal in the form of a rod, the *c*-axis being inclined at an angle of 60° to the axis of the rod. The *c*-axis is therefore at an angle of 60° to the direction of propagation in the slow wave structure, in the plane normal to the array of parallel conductors.

The isolator material used in the maser described here is yttrium iron garnet (*YIG*) in the form of flat discs which are supported in a polycrystalline alumina slab. By suitable choice of the thickness of the discs (determining the shape demagnetizing factors) it is arranged that ferrimagnetic resonance occurs in the discs at the same static magnetic field as is required by the ruby. The required demagnetizing factor is calculated from the Kittel resonance equation[(22)] which for flat discs mounted away from conducting walls may be written:

$$\omega = \gamma\left(H_0 - \frac{4\pi M_s}{2}(3N_z - 1)\right),$$

where H_0 is the applied magnetic field, N_z is the demagnetizing factor in the direction of H_0, and $4\pi M_s$ is the saturation magnetization of the garnet at the operating temperature. Yttrium iron garnet is one of the few ferrimagnetic materials which are suitable for this application as it still has a relatively narrow ferrimagnetic resonance line at liquid helium temperatures.

The slow wave structure is a comb having the dimensions shown in Fig. 18(*b*) and provides a slowing factor of 110. The assembly is shown in a cut-away view in Fig. 18(*a*).

The input signal is launched on to the comb structure through a coaxial line the centre conductor of which continues as the conductor *E* of Fig. 19. This conductor terminates in a short length of coaxial line *X* the length of which may be adjusted by means of the short circuiting plunger *P*. Impedance matching is achieved by adjusting the position of this plunger and varying the proximity of the conductor *E* to the first conductor of the comb proper. A similar arrangement is provided at the output end.

An advantage of slow wave structures of the iterated conductor type is that they can be introduced in a magnetic equipotential plane of a rectangular waveguide propagating, say, an H_{01} mode, without substantially perturbing this propagation. The structure of Fig. 18 will therefore propagate certain rectangular waveguide modes at frequencies above that corresponding to a free space wavelength of approximately 3 cm. In the present maser the pump energy propagates in such a waveguide mode and is introduced

(22) C. Kittel, *Phys. Rev.* **73**, 155 (1948).

into the end of the slow wave structure by a direct connection from rectangular waveguide, the broad face of the waveguide being parallel to the plane of the comb conductors. A pump power of about 30 mW is required to saturate the 1 → 4 transition.

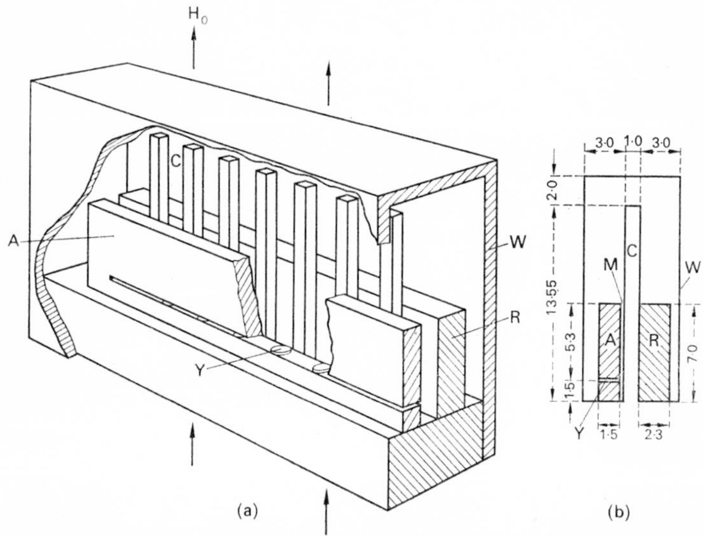

FIG. 18. Cut-away view (*a*) and cross section (*b*) of the 4170 Mc/s TWM. The dimensions are in millimeters. The slow wave comb structure *C* (pitch 2 mm) permits the propagation of waves with a group velocity of 1/110 times the free space velocity of light, and with an RF magnetic field circularly polarized in a plane perpendicular to the extension of the comb conductors. The sense of polarization in a travelling wave is opposite on the two sides of the comb, and opposite for forward and backward waves. In the applied static magnetic field H_0 of 3280 Oe the Cr^{3+} ions in the single crystal ruby slab *R* (Al_2O_3 with 0.05% Cr^{3+}) are resonant at the signal frequency (4170 Mc/s, 1–2 transition) and at the pump frequency (30 150 Mc/s, 1–4 transition). At the signal frequency they amplify forward waves. The yttrium iron garnet discs *Y*, held in place by the polycrystalline alumina slab *A*, and also resonant in H_0 at the signal frequency, attenuate backward waves. A sheet of 0.025 mm Melinex *M* is used to adjust the transmission characteristics of the slow wave structure. Pump power (30 150 Mc/s) is transmitted by the waveguide *W*.

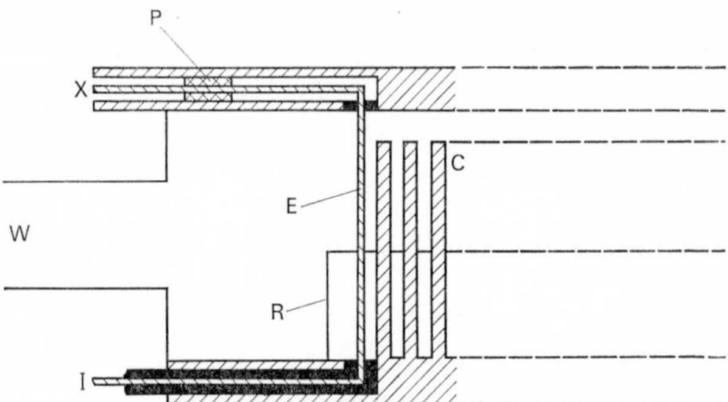

FIG. 19. Matching system from coaxial line to comb. *I* coaxial input line. *E* conductor, extension of the central conductor of both the input (*I*) and the terminating (*X*) coaxial line. *P* short circuiting plunger. *C* comb structure. *R* ruby slab. *W* pump waveguide. Matching is achieved by adjusting position of *P* and distance of *E* from comb.

The Cryostat

The maser structure is housed in the tail of a stainless steel double dewar vessel (Fig. 20), the magnetic field of 3280 Oe being provided by a permanent magnet as shown in Fig. 3. Centre frequency adjustment is obtained by varying the current through two coils attached to the pole faces of the permanent magnet. Connections from the top of the cryostat to the coaxial leads on the maser comb structure are made through low thermal conductivity coaxial lines which terminate at the top of the cryostat in waveguide-coaxial transitions (Fig. 21). The pump energy is introduced via a low thermal conductivity waveguide (WG 22).

To exclude air from the system (air is of course solid at liquid helium temperatures) it is necessary that all these leads should be vacuum tight. The input and output waveguides are sealed with thin terylene sheet windows and the pump waveguide with a mica window, all the seals being in the room temperature part of the apparatus. A series of carbon resistors attached at various levels to the connecting lead structure permits monitoring of the liquid helium level in the dewar vessel.

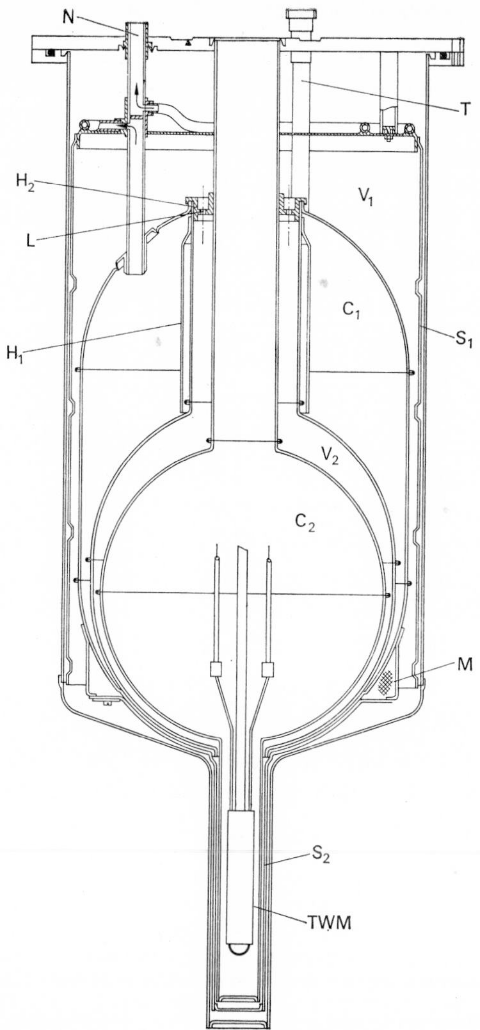

FIG. 20. Sectional drawing of double dewar vessel. Length 100 cm, max. diameter 33 cm.

C_1 liquid nitrogen container. C_2 liquid helium container. V_1, V_2 vacuum spaces. N nitrogen vapour outlet. S_1 copper radiation shield cooled at its top by nitrogen vapour. S_2 copper radiation shield cooled at its top by liquid nitrogen. H_1, H_2, copper heat conductors keeping a fixed point on the neck of the helium container at liquid nitrogen temperature. L holes linking vacuum spaces. T filling and venting tubes. M molecular sieve (getter material to absorb residual gases). TWM travelling wave maser.

The whole maser package is mounted on a cradle which allows it to be tipped through 45° from the vertical once it has been charged with liquid nitrogen and helium. This is its operating position when the aerial on which it is mounted is directed towards the horizon, and as the aerial moves to the zenith the maser moves through 90° to a position 45° the other side of the vertical.

The complete package is mounted on the aerial structure in its operating position, with a dewar vessel which allows an operating time per filling of liquid helium of about 8 hours. The maser is operated at a temperature of 1.5°K, the pressure over the boiling liquid being reduced by means of a pump situated lower on the aerial structure. Figure 22 shows a later version of the maser with a much larger dewar vessel which gives an operating time per filling of about 2 days.

The maser has been operated in a homogeneous magnetic field and also with a suitable stepped magnetic field to obtain greater bandwidth at the expense of gain. The performance figures of the maser under such conditions are:

Operation in a homogeneous magnetic field:

Electronic gain	52.5 dB
Bandwidth to 3 dB points	16 Mc/s
Total structure forward loss	11 dB
Total structure backward loss	70 dB
Net forward gain	41.5 dB
Noise temperature	15 ± 4°K

Operation in stepped magnetic field ($\Delta H \approx 5$ *Oe*):

Net forward gain	30 dB
Bandwidth to 3 dB points	28 Mc/s

These performance figures are in reasonable agreement with the design predictions outlined previously.

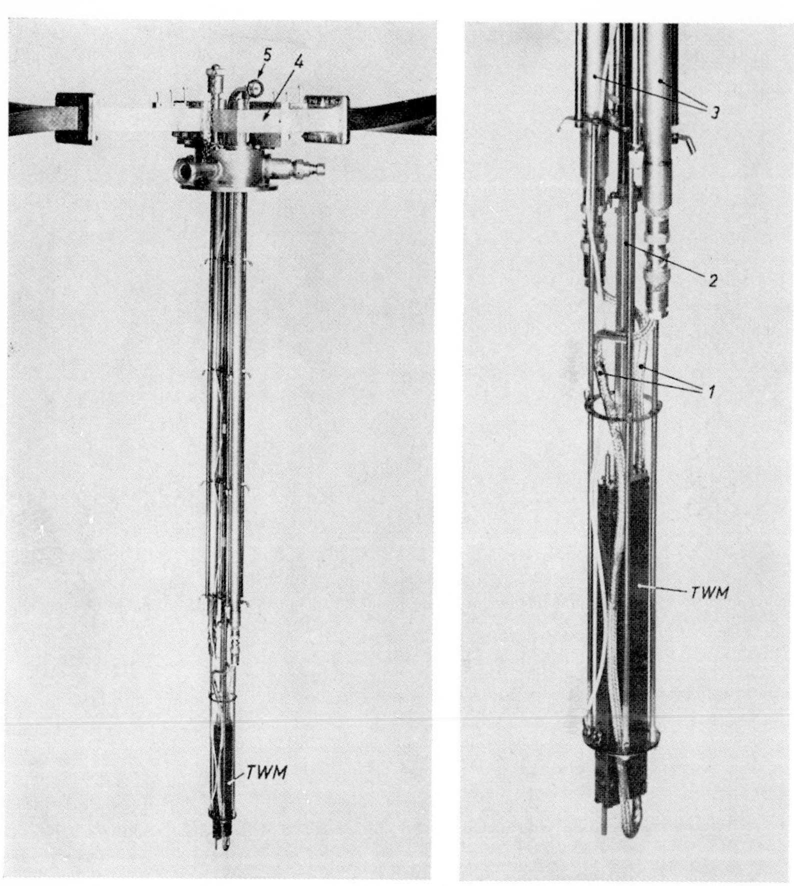

Fig. 21. Cryostat head and maser connection leads (enlarged view of lower end at right). *TWM* travelling wave maser. 1 coaxial leads (cf. fig. 19). 2 low thermal conductivity pump waveguide. 3 low thermal conductivity coaxial lines. 4 one of the two waveguide-coaxial transitions. 5 connection of pump waveguide.

Fig. 22. Travelling wave maser with large storage capacity dewar vessel. (Photo published by permission of the British G.P.O.)